D0169567

© 2001 by Michael Mazzeo

About the Author

IVAN SOLOTAROFF is a journalist
who has been published in *Esquire,*
the *Village Voice,* and *Philadelphia*
magazine, among other leading magazines. He is the author of a collection
of essays, *No Success Like Failure.* He
lives in Doylestown, Pennsylvania.

Also by Ivan Solotaroff

No Success Like Failure, Essays

THE
LAST FACE
YOU'LL
EVER
SEE

THE CULTURE OF DEATH ROW

IVAN SOLOTAROFF

Perennial
An Imprint of HarperCollins*Publishers*

Designed by The Book Design Group / Matt Perry Ratto

The Library of Congress has catalogued the hardcover edition as follows:

Solotaroff, Ivan.
The last face you'll ever see : the private life of the American death penalty / by Ivan Solotaroff.
p. cm.
Includes bibliographical references.
ISBN 0-06-017448-X
1. Executions and executioners—United States. 2. Capital punishment—United States. 3. Death row—United States. I. Title.
HV8696.S65 2001
364.66'0973—dc21   2001016604

ISBN 0-06-093103-5 (pbk.)

02 03 04 05 06 ❖/RRD 10 9 8 7 6 5 4 3 2 1

ACKNOWLEDGMENTS

FIRST AND LAST, to Robert Jones, who not only edited but substantially rerouted this book, several times, with subtlety and under difficult circumstances. And to my stepmother Shirley Fingerhood, who solved a big problem for me early on with professionalism and grace. To Jim Mairs, Gina Webster, and my father, Ted, for extreme generosity at a crucial time; and to my brothers, Paul, Jason, and Isaac, for support at the other crucial times. To Mike T., without whom this book could not have been written, and to Doug R., because of whom it almost didn't.

After four drafts, five years, and three full-time jobs, I'm sure I'm overlooking sources, interviewees, friends, colleagues, and employers to thank, beg indulgence from, and outright apologize to here. In alphabetical order, though beginning with Neil Fine, for a very special episode: Peter Alson, Bdog, David Boyer, Steve Bright, Celeste Byrne, Miriam Cabana, Alison Callahan, Andrea Casson, Mark Cohen, John Di Gennaro, Molly Friedrich, David Graves,

Elzbieta Goniszewska, Joel Griffiths, Sandy Hingston, Patty Hocutt, Gary Hoenig, Tsan Huan, Chung Huang, David J., Mark Jacobson, Mark Jenkinson, Eliot Kaplan, Jill Kearney, Patty Kelly, Ed Kosner, Barbara Kupidlo, Hugo Lindgren, David and Herb Lipson, Stephen McDonnell, Theresa Mays, Milesov, Albert Mobilio, Barbara Moss, Irene Mueller, Brendan O'Connor, Eric Perret, Robert Pierce, Sister Helen Prejean, Max Potter, Chris and Michelle Powell, Chris Raymond, Tony Reffuge, James Reed, Chris Schwartz, Daniel Shimberg, John Skipper, Pierre Vaz, Donnie Zittorelli, Jr. and everyone at Holt's, Rick Waechter, Ben G. Wallace, Craig Winston, and Richard Wormser.

To my subjects, particularly Don Cabana and Don Hocutt, who gave so freely of their time and candor. And to my wife, Janet Byrne, and my sons, Daniel and Zack.

CONTENTS

FEBRUARY 24, 1995. COLONEL DONALD HOCUTT, executioner for the State of Mississippi, pulls his Ford truck up to the Parchman State Penitentiary exit checkpoint off Highway 49W, halfway between Jackson, Mississippi, and Memphis, Tennessee. A guard stepping out of the checkpoint booth for a mandatory truck search sees Hocutt at the wheel and says "Sir" reverently, but he's barely acknowledged. Hocutt has been looking for words to describe to me the weird "crackling" that comes into his head every time he's in this prison farm now. When he concentrates, particularly on something that relates to his job, it's as though the world doesn't exist. "I could say it's like a noise," he tries. "Like a radio stuck in between stations. But it's really more of a feeling, like I was living under power lines or something. It's not pleasant."

This is one of the last times Hocutt will pass the checkpoint, after twenty years of corrections work. We drove up this morning for him to get some paperwork for his medical discharge stamped in the administration office, then he gave me a tour of Parchman's 19,000 acres, from the gas chamber to the warden's house. His paperwork, old-fashioned-looking forms of various sizes, has been in his lap the entire tour. He keeps touching it like a high-school senior with his last yearbook, squaring corners, adjusting paper clips, and smoothing carbons with an awkwardness that's very unlike him.

"It must be hard to be leaving after all this time," I say. "That's a whole world in there."

The guard has given the inside and back of the truck a once-over, and we're cleared to go. Hocutt opens the leatherette console between us and wedges his paperwork in between an old cassette of *Exile on Main Street* and a huge nickel-plate Colt .45. Then he turns south out of the exit on to 49W and floors it, staring ahead with one arm stiff at the wheel as the truck picks up speed down the flat, two-lane Delta highway. I can't take my eyes off the Colt.

"Donald," I say, touching the barrel. "That's a big gun."

"That's a dangerous gun," he says softly. "Maybe you don't want to be touching it." It turns out he keeps it loaded, with an eight-shot clip of Plus-P "cop-killer" hollow-point bullets. It's also cocked at all times, like his eight other pistols and rifles, with only the safety on. "I'll be damned if I'm going to die trying to get my gun loaded. Watch me close now." He takes the wheel in both hands.

I'm not sure what I'm supposed to be watching. Hocutt is a huge man, easily 300 pounds, with thick, baby-face features that cloud over dramatically when he concentrates or falls into one of his moods. I hear a rustle of paper in the glove compartment, then see he's gotten his right hand off the wheel and down into the console for the Colt without my noticing. He eases the gun out and across his lap, releasing the safety with his thumb as he raises the barrel under his left elbow and aims out the driver's window. "Pow," he says softly, releasing his breath.

"I do that a lot when I drive," he explains, sliding the safety back on as he returns the gun to the console. "You'd have never heard me just now, but for that paperwork."

"Why do you do that?"

"Just getting ready for the day someone tries to creep up on me."

"Where?"

"I don't know. In some parking lot."

"You have enemies from being the executioner?"

"None I know about. That's actually made me a pretty popular fellow around here."

Miles past the checkpoint, Hocutt nods over my head and says we're still driving past prison grounds. "Everything to your right, as far as you can see, is Parchman. We'll have guys escaping sometimes, and I'll catch them, two, three days later, deep into the woods. They look sad that they've been apprehended, but you really have to see their faces when I tell them they never even made it off the penitentiary grounds."

He points to the raised train bed of the old Illinois Central, ten feet past the road's right shoulder, and tells me about taking the train up those tracks when he was young to see his grandmother in Memphis. Over the raised bed, a half mile into the unfenced prison grounds, I can see the pink guard tower of Parchman's old maximum security unit. The smokestack of the gas chamber in the adjoining death house is just visible to the right of the tower. "When I was a kid on the train," he says, "I'd get a seat on the left and start looking out the window half an hour before we even got to that pink tower, just to make sure I didn't miss it. You can see, this isn't the most beautiful countryside, but it looked like something out of Marco Polo to me." When he came back from Memphis at night on the Midnight Special, he'd sit on the right and fight sleep, just to make sure he'd see the tower again, glimmering in the lights of the prison farm.

"I love the Parchman State Penitentiary," Hocutt says with sudden emotion, taking weight off the pedal. "And I have no regrets about giving it the twenty best years of my life. I just can't be there no more, that's all. To answer your question, though. It is kind of hard for me to leave."

I ask if the fate of Parchman's gas chamber is weighing on him. Hocutt's identification with it is strong, and the chamber has several parallels with his life and career. Installed in 1954, the year he was born, it was shut down in 1972, when the Supreme Court abolished capital punishment in America, then reopened when the Court reinstated the penalty in 1976, months after he began work at Parchman. A month ago, just as Hocutt's medical discharge from state service came through, the chamber was all but banned by the Jackson legislature, as part of Mississippi's eleven-year move toward lethal injection as its method of execution.

"It's just a relic," he shrugs. "Of the sixty-plus men on death row now, they're all due to get the needle, except for five who are grandfathered in for the gas. Two of them are severely retarded and a third changes personalities every day—Billy the Kid, Napoleon, Jack the Ripper. They've been talking about unbolting the chamber and putting it on display in Jackson, but no museum's shown any interest. Kind of surprises me. That chamber's got a lot of history to it.

"It's funny. People love the death penalty. They come up all the time, asking me about executions. Once I get started, though, they really don't want to hear *too* much about it. You know, I never heard of no kid telling his dad, *When I grow up, I want to be the executioner.* But when I was starting off as a Parchman guard, I couldn't wait for my shifts to come." Hocutt's face clouds over and he shakes his head. "I'm really not feeling too good today."

On the truck's cell phone, he hits the speed dial for the Memphis psychopharmacologist who prescribed a strong mood elevator for him a few days ago. It's not the first drug or doctor he's tried. At forty-two, Hocutt is "shot from the ground up": gout, maturity-onset diabetes, diverticulitis, arthritis in his upper body, partial deafness in one ear. His mind hasn't been right for years. Depressions steal over him, and for weeks he finds it almost impossible to get out of bed; the depths are followed by spurts of glee filled with plans and fantasy that keep him up at night. At the slightest provocation, he falls into rages and incessant replays of some injustice, violence, or close call from his two decades on the job, and he broods endlessly about the fight he's waged with the State of Mississippi for the past three years to get a full medical discharge. The net result is morbid hatred—"a constant negative draw," he sometimes calls it. "Like I'm on a planet where gravity is five times denser than on Earth."

The nurse who takes his call in Memphis is the one he likes. "What's it going to take before I feel better?" he asks plaintively, his voice up half an octave.

"Your new medication's operating on a different part of the brain," she assures him over the speakerphone. I can hear why he likes her. Her voice, a gravelly smoker's voice, is soothing in its lack of affect—you can tell she isn't bullshitting—and her accent is strong the way Hocutt likes it: not just southern, but deep country. When she says *brain*, it sounds like *brine*. "It might take another two weeks for it to become fully thair-apeutic. But you know that, Donald. What else is bothering you, hon?"

"I'm fine." Hocutt says nothing for a second or two, then thanks her before hanging up.

"It's probably just the thought of another day without a decent meal," he says a mile down the road. His various medicines, in combination, wreak havoc on his stomach, and his diet is limited and alien to him. "There's also all this fat to deal with," he says. "I've always been big, but never like this." Several doctors have told Hocutt that his weight makes his ailments, severe in themselves, potentially life-threatening. "I really don't care if I live or die," he says—a refrain so familiar I tend to tune it out, until I remind myself that this is an executioner, talking about living under a death sentence. "I figure I've got three or four years left, any way you look at it. I just want to be around to see Mark [his ten-year-old] play football. So I take the pills and follow these diets. For the most part." The latest has him down to a glass of skim milk, two bananas, an oatmeal cookie, and a dozen pills for breakfast. The rest of the day is at the mercy of mood swings and the medications' side effects. Sudden rages or fits of gloom send him straight to the fridge or the nearest Sonic or Checkers, as though he could eat his way out. The diuretics he takes for gout make him urinate frequently, throwing his electrolytes off. To restore the lost potassium, he eats bananas, a half dozen at a sitting sometimes, and they bind him up terribly.

"Fuck it," he says, hitting the speed dial for a greasy spoon two miles ahead. He tells the woman who answers it's Colonel Hocutt calling, and to have a burger and a large fries ready. "Sound good?" he asks me.

"Sure."

"Make it two. I'll take mine all the way."

"Tell me about it," the woman says.

"I like mine piled with fixings and dressing," he explains sheepishly when she hangs up. "Piled so high the grease goes down your elbow when you bite into the burger. Do we really have to go to this graveyard?"

Hocutt has offered to take me on a tour of Jimmy Lee Gray's grave outside Indianola, half an hour down 49W. Several times this morning, he's tried to talk me out of going, then just as quickly talked himself back into it. Twelve years have passed since Jimmy Lee Gray was executed in Parchman's gas chamber, but people keep asking for a tour of the grave from Hocutt, who mixed the sulfuric acid bath that dissolved the cyanide that killed Gray. He usually takes the people who ask, though he says there's nothing to see. "They always spend a lot of time by his stone, for reasons I cannot fathom," he tells me. "No one except the lawyers, church ladies, and the death penalty weirdos took any interest in Jimmy Lee Gray while he was alive. People on death row get a lot of mail, you'd be surprised, but he didn't. His family never visited, and his victims' families didn't show up for his asphyxiation. Even the guys on the row didn't want to know nothing about him, and they're so low they need binoculars to see a centipede shit."

Gray was famous enough afterward. He took a long time to die, or so most of the witnesses to his asphyxiation claimed. A doctor monitoring his heartbeat on a remote EKG said he was clinically dead two minutes into the execution, but his head banged back repeatedly on a ventilation pipe behind the chair in the chamber, and the legend has grown that his death was a ten-minute torture.

It was the first time in nineteen years that the gas chamber had been used, and it was the beginning of the end for it. By turns, Hocutt says he believes the doctor's assessment of a quick death and that he truly doesn't care.

"I could have sat on Jimmy Lee and ate a sandwich while he asphyxiated," he tells me. "It wasn't pretty, and the appearance of it having been bungled reflected badly on the State of Mississippi, but Jimmy Lee Gray was a piece of shit. He took a three-year-old girl into a logging woods outside Pascagoula, raped her, sodomized her, pushed her face into a bog by a drainage ditch until she choked on mud, then threw her body off a bridge into a stream. As far as I'm concerned, Jimmy Lee Gray died too painlessly. He's really the only one I wish they could have done differently."

"How?"

"That's legal?"

A cousin of Hocutt's, a massive man named Red, is at a window table in the greasy spoon when we pull up. They talk about catfish farms and last year's soybean crop and gossip about people they recognize driving by on 49W while we eat. Hocutt and I plow through two-inch-tall burgers while Red carves mouthfuls off a dainty square of gray meat in butcher paper with a penknife. I ask if it's pâté he's eating.

"We call it sow," says Red.

"Pig liver?"

"I think they grind the whole animal up." He looks at it. "Just real concentrated, I guess. What brings you down?"

I tell him we're headed to see Jimmy Lee Gray's grave.

The cousins exchange a glance over their food. "What do you think, Red?" Hocutt asks with a deadpan look. "Think he's still dead?"

Red wipes both sides of his penknife on the butcher paper and folds up the remainder of his meat square. "I guess you got him pretty good, Donald."

"Guess we did at that." Hocutt cleans grease off his lips with a napkin, and throws it like a white towel over the last quarter of his burger. "You know, I really don't mind going out to that grave-yard," he says. "Fire ants are thick up there, but they're no prob-lem if you know where not to step. It's pretty peaceful, actually."

The graveyard is on one of the delta's rare promontories, off a winding, rutted blacktop up from a cypress grove. Hocutt points to where Jimmy Lee Gray lies, says to take as long as I like, then picks up a stick, walks to the end of the yard, and settles down cross-legged with his back against a huge tree. Gray's stone is a generic red slab etched with his name, dates, and some minimal scrollwork. I spend a long time in front of it, mostly looking at the final date, September 2, 1983. "Now it is a terrible business," wrote G. K. Chesterton, "to mark out a man for the vengeance of man." I feel that strongly, looking at the ordained date of Jimmy Lee Gray's death: the *terrible business* of *the vengeance of man.*

"I want a simple stone for myself," Hocutt says when I join him under the tree. "The classical shape with the round top and the beveled edges. They just get better-looking as time goes by." His face looks like an old tombstone: heavy, impassive, full of resigna-tion and imponderable meaning. With the stick, he points out stones that are centuries old, lingering with approval on a few that are so deep-sunk and weathered you can't read their inscriptions. "There'll be nothing on mine but the name and dates. Maybe a lit-tle something like, *He raised two sons and now it's time to rest a spell.*"

"Do you think much about an afterlife?"

"I believe that life on earth is hell."

He jabs the ground with the end of his stick. "I have thought about it. You can't put a man to death and not have it cross your mind. *Where's this guy going from here?* I have no idea if there's an afterlife, but if there is one, right here is pretty much how I see it for myself: sitting back against an old tree for a long time, wip-ing your brow, and just saying, *Phew. I made it.*"

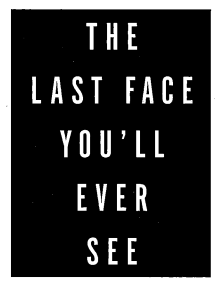

THE
LAST FACE
YOU'LL
EVER
SEE

THE "PECULIAR INSTITUTION" OF AMERICAN EXECUTIONS

IN 1995, WHEN I BEGAN *THE LAST FACE YOU'LL EVER SEE*, 77 percent of Americans favored the death penalty. As I redraft this introduction in late 2000, the Harris Poll shows that sentiment has dropped to 66 percent; subsequent and more extensive polls show the numbers softening further. The radical shift of opinion is generally attributed to the recent suspension of Illinois executions by Governor George Ryan, a pro-penalty man who'd been moved by proof of widespread evidence of innocence on his state's death row.

On my desk is a clipping from yesterday's *New York Times* about a Supreme Court stay of execution of a severely retarded Texas man—a story that snagged front-page space from the postelection fight over recounting Florida ballots for the U.S. presidency; buried in today's paper is a brief wire item about last night's lethal injection in Texas (of a different condemned man), which brought that state within two executions of a record year. A poll in yesterday's *Times* showed 60 percent of Americans favoring a recount of the Florida ballots; on the radio this morning, I hear that number has dropped to 45 percent. "Americans want closure," said the news anchor, adding, a few sentences later, "You're watching democracy in action."

It is against this backdrop—of mass opinion swaying on whether someone should live or die, of democracy as a daily PR battle between opinions posing as truth, fact, or law—that this book about executioners makes most sense. Other than abortion, there is simply no issue on which opinions are more divided, deeply felt, and yet inconstant, or where contentiousness blurs the reality of the event so strongly. Thus, I focus on the coup de grâce itself, taking a mandate from two sentences in Albert Camus's "Meditations on the Guillotine": "[We] must show the executioner's hands each time and force everyone to look at them. . . . Otherwise society admits that it kills without knowing what it is saying or doing."

I undertook this imperative not as some existential errand but out of curiosity. I wanted to know who carried out executions, where, when, and how, and what it had been like for them to do it. As I gathered details, however, I saw that these concerns addressed only the *doing* of Camus's sentences, not the *saying,* and that brought me to a further question, both simpler and more difficult to answer: *Why* did they do it? Or more exactly, Why do we *want* to do it?

Though even asking such a question implies the second, moral question—"Is the death penalty right or wrong?"—I have made no attempt to answer that in this book. I wouldn't ask it of an executioner to begin with, and after living with the complexities of executions for five years while researching and writing this book, I feel only more agnostic on the issue than when I began. Instead, I have tried to simplify my "deep structure" by asking only about the *motive* of capital punishment, much as a death penalty statute asks after the motive of a capital crime. Is an execution a rational mechanism— i.e., a tool of deterrence, punishment, or jurisprudence by which a certain type of murderer is, in John Stuart Mill's words, "solemnly blotted out from the fellowship of mankind"? Or is it something altogether different—an expression of an irrational urge far more subterranean than the will to justice?

Americans rarely ask such a question, because generally they know the answer. They do something because they want to, and because they can. A contemporary American execution, scorned by abolitionists abroad and at home as a form of moral backwardness, is probably nothing of the kind. Like mid-nineteenth-century slavery, it is rather our "peculiar institution"—and as anyone who has toured a death row or attended an execution can attest, the will to enslave and the will to execute are either the same or remarkably similar. The comparison first suggested itself to me in the Deep South's death houses, which sit in prison farms culled from former plantations, but it was reinforced, time and again, in the abjection I've seen on the faces of men on death rows across America—a beaten, naked, animal look that one sees nowhere else. Whether one is for or against the penalty, there is simply no denying that an execution is an extreme act of subjection.

While almost every nation we would compare ourselves to has given up killing its citizens, calling it a "primitive" or "barbarous" act, an execution is something we are able to continue to do not because we are more primitive or barbarous, but because we have adapted. All twentieth-century European executioners (save the occasional military-minded firing squad) killed in the primal manner of their predecessors—by attacking the neck, with a rope, guillotine, garrote, ax, or saber. America spent the century "civilizing" executions, through innovation: with the electric chair, an attempt by the New York State Legislature of 1890 to move past the brutality of the rope; with the gas chamber, used first in Nevada in 1924 when abhorrence for the gallows had led to a moratorium on the penalty; and with the lethal-injection gurney.

All three forms expressed the national will to get things done. It is probably no coincidence that Thomas Edison, *ne plus ultra* of American "can-do," was intimately involved with the first five electrocutions, or that a leading East Coast toxicologist who had seen the effects of German gas attacks in World War I came up with the idea of "canning" prussic gas in a chamber. Methods for humanely dispatching the condemned with lethal drugs had been considered by state legislators since the nineteenth century, but it took what has become known as the unofficial moratorium on capital punishment—the years 1967–77, in which no American was executed—to finally effect the adoption of the needle.

That unprecedented ten-year span, not coincidentally the only years in which a majority of Americans were consistently against the penalty, represents more than a hiatus in executions. Like much that happened in the sixties, it was the end of something—in this case, a mass will/desire to legally kill. That has clearly come back, but its expression is utterly changed. Compare today's somber, self-searching execution coverage to the "Gwoin to Heaven"–style headlines of 1933, the previous peak in frequency and popularity of the penalty. Executions were a kind of public burning even then. Conducted largely inside prison walls, they caught the homicidal passion of the masses in a way that would be unthinkable today. The executioner was still a public figure, and he had to be. He expressed a desire that was still very public.

All American executioners except one retired or died during the unofficial moratorium. The exception, whom we'll meet in Chapter 3, was a Mississippian named Thomas Berry Bruce, who had been executing since 1951. The link he provides—both to current executioners and the underlying motives of the penalty—is the first of two reasons I chose to situate the bulk of this book in Mississippi.

In 1972, only 40 percent of Americans polled by the Gallup organization favored executions. A month after the poll was issued, the Supreme Court abolished every capital punishment law in the nation, finding, in *Furman* v. *Georgia*, that the penalty had been wielded inhumanely and arbitrarily, violations of the Eighth Amendment protection against cruel and unusual punishment. It was a momentous decision, ending a practice that had taken the lives of some 18,000 Americans, and forever sparing 629 then on death row.

As the law of the land, however, *Furman* lasted less than five years. A 5–4 decision in which all nine justices wrote individual

opinions, several of which varied radically in tone and stance, it
was the longest judgment the Court ever handed down. Essentially,
it was less a mandate against capital punishment than an indecisive
text from which future legal debate on the death penalty would
spring: the punishment hadn't been wielded for six years, and there
were no serious execution dates impending, so the decision wasn't
stopping anything. Rather, its length and divisiveness provided
states largely with guidance in revising statutes. This was hardly
unnoticed by pro-penalty theorists, state legislators, and district
attorneys from Florida to California, states that combed *Furman*'s
nine opinions, overhauled capital statutes, and within a year had
begun to prosecute death sentences again.

In 1976, as opinion in favor of executions rose to 66 percent, the
Court upheld Georgia's most recently revised capital statute, which
had been the basis of a death sentence meted out in *Gregg v.
Georgia*. Along with guidelines gleaned from two other death sentences the Court had thrown out earlier in 1976—as coming from
revised statutes that were still "unduly harsh," "unworkably rigid,"
or below the "evolving standards of decency that mark the progress
of a maturing society"—the reaffirmation with *Gregg* paved the
way for any state that wished to resume executing.

There were many. Before *Furman*, most of the thirty-two states
that still executed had obsolete death penalty laws, some dating
from territorial days. By the time *Gregg* came down, thirty-five
state legislatures were debating, drafting, or revising capital punishment laws, and a new 600-plus men and women sat on death
rows across the country. After the opinionated ambivalence of
Furman—"a badly orchestrated opera," as one critic put it, "with
nine characters taking turns to offer their own arias"—the mandate
offered by *Gregg* and the other 1976 Court decisions was highly
pragmatic: Decide for yourselves who, why, where, when, and how
you convict, imprison, and execute, but avoid any impression that
the penalty has been sought, arrived at, or inflicted in a manner that
seems unduly harsh, unworkably rigid, indecent, etc. In short, let no
one be able to say your actions were arbitrary or inhumane.

As state legislatures edited post-*Gregg* statutes to accommodate the who, why, where, and when of this new mandate, wardens of the thirty-five prisons housing death chambers had to answer the how. It wasn't easy. Nine years had now passed since the last American had been executed, and in 1976 it was all but impossible to find gas-chamber and electric-chair experts. Missouri's chamber leaked. No one in North Carolina knew how to turn theirs on. Nevada's two-seated chamber in the state prison at Carson City had become the guards' pinball room, and its ventilation stack sagged. South Carolina needed a new cap for the head electrode of its electric chair and no one knew where to get one. Officials in Tennessee couldn't even tell the front from the back of their headcap. In Angola State Prison, shortly after Louisiana's revised capital punishment statute passed on appeal to the Supreme Court, Warden John Blackburn took the massive black electric chair out of storage in the garage of the warden's mansion. His wife had asked several times during the moratorium if she could paint it white, cut a hole in the seat, and use it as a planter, but Blackburn had a thing for electric chairs. He kept a three-inch replica on his desk that gave anyone who touched it a small shock, and he roared with laughter every time someone fell for it.

Florida's wardens, prior to executing John Spenkelink on May 25, 1979 (America's first electrocution in thirteen years), scoured the nation for electrocution experts, found none, and simply drew up the best protocol they could. They knew their chair used a two-minute cycle of varying voltages, but they had also been told by old-timers that it was never allowed to run that long. An abbreviated cycle was used, and it took three separate attempts to kill Spenkelink. Four years later, when Alabama electrocuted for the first time post-*Gregg*, the jacks connecting the chair to the generator were so poorly wired that three separate cycles were needed to kill John Louis Evans III, who suffered for twenty minutes while his lawyer screamed, "This is cruel and unusual punishment!" Governor George Wallace, reached by phone halfway through, told the executioner, Warden Charlie Jones, to continue until the sentence was carried out.

But writing protocols and equipping old chairs and chambers, as post-*Gregg* executioners learned, was only the beginning of their burden. The world had changed, "standards of decency" had indeed evolved, and what had been seen as an ugly but necessary job now seemed "unduly harsh" to many. In 1979, the anthropologist Colin Turnbull, allowed to tour Virginia's death row, in a basement of the Spring Street Maximum Security Prison in Richmond, got a sense of the antique gruesomeness of an electrocution, which would remain the most common form of American execution until it was overtaken by lethal injection in 1991. Off the death chamber, Turnbull found a "sandbag room," where the contorted corpses taken from the chair had been manually broken back into shape over a period of hours by weight of the bags.

The violent spectacle of electrocutions surprised and baffled the new executioners as well. When Robert Wayne Williams was put to death in Angola's chair in 1983, Louisiana's first execution in nineteen years, wardens were amazed that the chair literally cooked Williams's scalp and legs, which smoked and sizzled for several minutes. Twenty-four hours later, the corpse reeked so strongly that mourners found it difficult to remain in the funeral parlor where Williams was laid out. Angola's warden, Ross Maggio, had to call sources outside the prison simply to learn if this was the way it was supposed to happen.

The gas chamber was harder still for many to watch in action. The apparent violence of asphyxiations grievously offended post-*Gregg* witnesses in state after state, from the 1983 execution of Jimmy Lee Gray in Mississsippi to the gassing of Donald Eugene Harding, on April 11, 1992, Arizona's first execution in thirty years. A Tucson television reporter sobbed uncontrollably during Harding's ten-minute execution; two other reporters "were rendered walking 'vegetables' for days"; the attorney general vomited halfway through; a prison staff member who ran the execution likened it to watching a man suffer a series of heart attacks; and the prison's pro-penalty warden said he'd resign if the state told him to run another asphyxiation. But Harding's death was probably no different from those suffered in Arizona's chamber since it was installed as "a humane measure" in 1933, replacing a gallows that had decapitated a condemned woman.

The difference was really only one of perception, but over time it was big enough to put an end to these old methods. Contemporary distaste for asphyxiations was most apparent (and documented) with the famous "Green Room" in the basement of California's San Quentin Prison. The nation's busiest chamber for thirty-five years before *Furman*, it saw only two post-*Gregg* asphyxiations, though the state's death row population was America's largest for much of the last quarter of the twentieth century. U.S. District Court Judge Marilyn Hall Patel, faced with ACLU lawsuits challenging the chamber as cruel and unusual punishment, ordered a videotape made of the 1992 asphyxiation of Robert Alton Harris, then studied detailed medical records of past asphyxiations, depositions of Holocaust survivors, and expert testimony on cyanide poisoning. She banished the chamber two years later, declaring Harris's asphyxiation "comparable to disembowelment, drawing and quartering, and burning at the stake." Compare this to earlier language, of the California Supreme Court, in *People* v. *Daugherty* (1953). That court, which had examined fifteen years of medical records, said the chamber met "contemporary scientific standards. . . . For many years," it reasoned, "animals have been put to death painlessly by the administration of poison gas."

The two courts were separated by forty years, but they looked at the same act. Medical records show that Robert Alton Harris's ten-minute death in 1992 was similar to what had been suffered by the condemned since December 1938, when California switched from the gallows to gas. The chamber was in good condition in 1992, the old executioners had left a nine-page protocol for its operation, and the wardens even had a scrapbook, offered by Joe Ferretti, the retired, diminutive, seventy-nine-year-old guard who had served as a death watch "baby-sitter," an officer who sits out the final hours with the condemned. A litany of last nights, words, meals, and idiosyncracies of the 129 executions Feretti had worked, the scrapbook was most telling in what it omitted: that the condemned appeared to strangle for anywhere from one to seven minutes before losing consciousness. For Ferretti, by several accounts a "solicitous," "sensitive" man who had become intimate with the condemned in their last hours, that apparent suffering wasn't noteworthy. An open secret, perhaps a slight embarrassment, it was *implicit* in the penalty, as indeed it always had been.

After *Furman,* however, the spectacle, legacy, or threat of the violence of the chair and chamber made it progressively difficult for states without lethal injection to execute. Die-hards like Alabama, Arkansas, California, the Carolinas, Colorado, Florida, Georgia, Indiana, Kentucky, Maryland, Mississippi, Nebraska, Ohio, Tennessee, Virginia, and Washington fought through the 1980s and 1990s to retain their old methods, which, somehow, continued to meet the "evolving standards" of the Supreme Court. In 1996, for example, the Court set aside Judge Patel's ruling, briefly reopening California's chamber, and throughout the 1990s it upheld the constitutionality of Florida's chair—the nation's busiest—despite a spate of horrific electrocutions. Instead, it was pro-penalty state legislators and governors who gave in, frustrated by the snail's pace caused by ongoing legal challenges to the old methods. The Arizona House of Representatives voted to banish its chamber and switched to lethal injection three weeks after Donald Harding's gruesome death. North Carolina closed its chamber after the 1994 asphyxiation of David Lawson, who went to his death screaming, "I'm human, I'm human." Maryland's chamber was outlawed in 1997, Mississippi's in 1998. Before *Furman,* there were eleven gas chambers operating—all have since been closed. As of this writing, only Alabama continues to electrocute with any frequency.

There may well be a similar argument, two or three decades hence, about the humaneness of lethal injections. The $70 of chemicals that effect death, known among corrections professionals as Texas Mix, or Texas Tea, are kept in separate drip bags on an IV stand in an unseen executioner's room. The drip bags are connected to one or two central tubes, which are run out to the death chamber and hooked to a catheter attached to a needle that has been inserted in the arm of the condemned. The three drugs are injected in stages, with a wash of distilled water run between each through the tubes (and the condemned), as the poisons clot and throw off a viscous white precipitate when combined.

The first drug is either sodium thiopental or sodium pentothal, which slackens the muscles, depresses the central nervous system, and induces something resembling sleep; then pancuronium bromide or Pavulon (artificial drugs similar to curare, used by Amerindians to poison arrow tips) is administered, to attack the lungs; and finally, potassium chloride, an electrolyte that, in large doses, reverses the polarity of the heart muscles, causing failure. All three are used in common medical practice—the sodium compounds are still used in dental anesthesia; curare, greatly diluted, of course, is a homeopathic remedy for anxiety; potassium chloride is prescribed for heart fibrillation. This quasi-medicinality enables the common opinion that lethal injections are like "putting people to sleep," a homespun wisdom first offered in 1973 by then California governor Ronald Reagan, who cited his years as "a farmer and horse raiser" in making the assertion.

Consciousness is a relative thing, however, particularly when large amounts of strong drugs are involved. Thiopental and pentothal certainly make the condemned appear to lose consciousness, but there is no way of measuring this: all three parts of the lethal cocktail are of a sufficient dose to cause death by themselves. The sodium compounds, however, may only render the condemned incapable of expressing the pain and panic that a poison like Pavulon would cause anyone even remotely conscious, or the subsequent terror when the heart's rhythm is suddenly altered. If so, death by lethal injection may not be the euthanasia of common perception but a chemical entombment, lasting anywhere from two to ten minutes.

Statistically speaking, however, the gurney's medical image has gone a long way toward anesthetizing us to an execution's reality.

Pro-penalty sentiment reached its peak in the years 1989–94, topping out at 79 percent, exactly as electric chairs and gas chambers were closing down and the needle became the abiding metaphor of American executions. Hence, the second reason this book is weighted toward the four Mississippi executions—they happened in a gas chamber. Short of a botched hanging, it is the most unattractive method this country has to offer, the most onerous on the men who carry it out, and therefore the best locus for examining the deeper motive behind an execution: a contemporary executioner could not do more than one without having his resolve (and motive) tested. As the reader will see, asphyxiations drove one executioner out of a lifetime career in corrections, and another deep into the hell of this "terrible business [of] the vengeance of man" that I set out to explore.

The gas chamber also sorely tested the will to execute of one of America's more stridently pro-penalty populations. The four gassings chronicled in this book were the only executions Mississippi managed to carry out in the last quarter of the century, despite the fact that the state sits in the middle of what's known as the Death Belt, the 900-mile swath across the Deep South that, since 1976, has been the heart of capital punishment in America.

CHAPTER ONE
THE DEATH BELT: FLORIDA TO TEXAS

THERE WERE 358 EXECUTIONS between 1976 and 1996, the first twenty years after the death penalty returned to America. Sixty percent were carried out within a two-hour drive of Interstate 10, the southernmost of the coast-to-coast highways. The five eastern states of I-10 are called the Death Belt because of the disproportionate number of executions carried out in their prison farms: in "Old Sparky," Florida's three-legged brown electric chair, which sits in a small, bright room in the state prison at Starke; in "Yellow Mama," the chubby yellow electric chair at Holman State Prison in Atmore, Alabama; in "Black Death," the metal chair in Parchman State Penitentiary's silver gas chamber; in Louisiana's lime-green death house, five miles deep into the woods of Angola, an hour north of Baton Rouge; on the lethal-injection gurney in the midnight-blue death chamber of the "Walls Unit," the massive brick prison that occupies much of downtown Huntsville, Texas. Run a finger down a list of the condemned and notice the three-part names: John

Louis Evans III, Jimmy Lee Gray, Robert Wayne Williams, James Dupree Henry, Alpha Otis Stevens, Robert Lee Willie, Edward Earl Johnson, Connie Ray Evans, Arthur Lee Jones, Andrew Lee Jones. That's when you know you're in the Death Belt.

A large, unpainted wood-frame house in Headland, Alabama, fifty miles from I-10, is a de rigueur first stop on any trip through the Death Belt. The ancestral home of Watt Espy, America's fore-most historian of executions, it also serves as the offices of his Capital Punishment Research Project. An unfunded, unaffiliated, one-man attempt to collect every available fact about the American death penalty, it is a project to which he has devoted most of his adult life.

Espy, who greets me at the door, is a tall, thin man in black frame glasses, a polo shirt, khaki pants, and tennis shoes with prim anklet socks. He smiles a lot when he speaks, heavy, preoccupied smiles filled with a mournful irony. "These are the condemned," he says, pointing to the head shots, mug shots, wire photos, and book and magazine clippings, all individually framed, that fill the walls. Many hang at derelict angles, others have a crack in the frame glass, and a few show the yellow discoloration that comes with prints taken too soon from the hypo bath, but there's no mistaking the passion with which they were assembled.

Espy tells me their stories, one at a time: a man who willed him-self into a coma and had to be carried to the chair; another who strolled in blithely, saying, "I'd ruther be fishin'"; one who came in with a cigar and a pink flower in his buttonhole; a man who had printed the prison tattoo H-A-R-D L-U-C-K on his knuckles; another who handed the electrocutioner a check for his $150 fee, signed "The Devil"; one who asked for bicarbonate of soda before entering the gas chamber; one who said the soup of his last meal was too hot; one who complained from the electric chair, "I sick. I eat too much." Ones who read verse: "Hang me high/And stretch me wide/So the world can see/How free I died," or quoted rap: "You can be a king/or a street-sweeper./But everyone gotta dance/with the Grim Reaper." Those who told their executioners,

"Step on the gas"; or "I came here to die, not to talk"; "I am Jesus Christ"; "Hurry it up, you Hoosier bastard"; a woman who warned, "My blood will burn holes in their bodies."

A thin strip of wall in Espy's office bears the house's only images of the living. Six are of his family, including three of an older brother who runs the local savings bank. The other is a head shot of Mario Cuomo, who vetoed every capital punishment statute drafted by the New York State Legislature in his twelve years as governor. "As you might guess," Espy boasts, "I'm a bit of an abolitionist myself." He's also proud to be a recovering alcoholic. "I stayed drunk for the only execution that I will ever attend. It made me physically ill. I vomited."

"Which was that?"

"That would be the botched electrocution of John Louis Evans the Third, State of Alabama, April twenty-second, nineteen eighty-three."

The office, once the living room, has a cranky old copier, a low-end computer, a magnificent library of books on the death penalty, a few dozen photos of the condemned, and a row of file cabinets from the Dewey decimal era, which Espy uses to store execution quotes, names, and factoids on four- by five-inch index cards. The large room is otherwise given entirely to shelves stacked with loose-leaf binders formerly used by traveling salesmen. The binders, which Espy buys for a dime apiece in close-out sales, are mystifying at first, as they have their original logos on the spines and covers—everything from Coca-Cola to local hardware companies. Inside each, Espy tells me, are the details of an era's executions, arranged alphabetically state by state. "That's Indiana behind you"—he points to a half dozen binders—"then over into Idaho, Kansas on the far side, down into Louisiana, Mississippi, up through Nevada, Pennsylvania, Ohio, and down again into Texas, which is taking up its share of my attention these days."

I tell Espy I'm headed to Huntsville to see a condemned man, Noble Mays, with whom I'd exchanged letters, then to Louisiana to attend the lethal injection of a man named Antonio James. Espy asks how I feel about corresponding with a condemned man.

"It's a little creepy," I admit.

"I've learned to avoid all contact, though they write me often." Espy seems to have mixed feelings about my attending an execution as well, though his eyes light up when I mention I'll be going through New Orleans. "I haven't been there for ages," he says. For the first time, I notice the room has no windows. The house is extremely dark.

Espy has supported the project with odd jobs—selling cemetery plots or the *Encyclopaedia Britannica* door-to-door. Once a year, he borrows against stocks held in trust, and he says he's having a hard time keeping the project going. At the time we meet, he has chronicled the details of 18,812 American executions—by hanging, shooting, electrocution, gassing, lethal injection, burning, beheading, entombment, gibbeting, breaking on the wheel, boiling in oil, roasting, drowning, etc.—more than 10,000 of which were unknown to posterity before he tracked them down. *Uncollected* is the word Espy prefers. "Virginia leads by far," he says. "Two thousand and forty-nine thus far."

"Why Virginia?"

"Because they executed so many slaves. The closer you get to the Mason-Dixon Line, the more perilous it's been, historically, to be of color. My records show that clearly."

Truly grotesque executions tended toward the north of the Mason-Dixon Line, Espy emphasizes, though he has collected a fair. share of Death Belt horror stories. He waxes particularly eloquent on Louisiana, where slaves and mutineers were nailed into boxes and then sawed in half, roasted on cannon barrels, or sewn into a leather sack with a dog, viper, monkey, and cock and thrown into the river. "But the worst was of a slave in colonial New York who poisoned her master. Spit-roasted on a slow fire for hours and hours"—Espy gets up to enter a name in a massive accountant's ledger on an old lectern—"with a horn full of cool water held inches from her lips, to accentuate her agony. In the modern era, we've lost sight of the fact that executions were meant to inflict not only death but pain.

"Make that eighteen thousand, eight hundred, and *thirteen*." He inscribes the man's name, speaking phonetically as he writes: "Nelson Shelton, Midnight, March the seventeenth, nineteen hundred ninety-five. State of Maryland. I've been busy all morning and forgot to put his name in. It was going to be eighteen thousand, eight hundred fourteen, in a double execution, rare for the State of Maryland, but his brother received a stay, so he could leave a kidney to their ailing mother." He returns to his desk. "And he doesn't have long, does he?"

"Is that your Book of the Dead?"

"I just call it Ledger Number One. So tell me what brings you down here."

I'm ten words into quoting Camus's sentences when Espy smiles by way of interruption. "There's a problem with that very beautiful quote," he says. "Camus meant it as a challenge, of course, but society actually kills best and most frequently when it knows *exactly* what it is saying and doing. If my research bears out any one truth, it's that."

As Espy speaks, I remember some difficulty I'd had verifying a story Charles Manson told of his mother witnessing a hanging in the late 1930s in Kentucky's Moundsville Prison, when he was four years old. In Manson's telling, his mother had been hiding near the gallows, trying to avoid her work detail, when the hanged man's head was ripped off by the fall from the gallows and rolled past her. "That would be the Hyer case," Espy says without blinking. "And given security conditions in Moundsville in that period, yes, it would have been possible for her to have witnessed an execution she had no right to. And who knows but what effect seeing that had on her, and eventually had on her four-year-old progeny? See, I believe that more than one person dies with each execution. They exert a horror and a fascination that never really goes away."

Espy dates his fascination with executions to a UPI wire sent out on the morning of June 20, 1953, when he was the teletype operator for the U.S. Navy base in French Morocco. The subject was the Rosenbergs' electrocution, and much of the crew was waiting outside his office for the message. "Six bells from the teletype," Espy remembers. "Then it came: ETHEL HAS HOT PANTS. It was just some wire stringer's puerile joke, but I'll never forget it."

Years before evidence emerged that Julius Rosenberg had in fact delivered military secrets to the Russians, Espy had investigated the case and come to the decision that Julius, and perhaps also Ethel, didn't belong in his file of *Wrongful Executions—Confirmed Innocents*. "That was a painful conclusion," Espy says, handing me the file, "because their deaths put me on my life's work." It's that honesty, however, that enables Espy to earn an occasional, much-needed fee as an expert witness in capital punishment trials. As every lawyer learns from Espy, the difficulty in assembling accurate execution data comes from the mixture of sensationalism and obscurity that surrounds the practice. "Executions tend toward glaring headlines," he says, "but the details have a way of being omitted, gotten wrong, misinterpreted, hidden, whatever."

An abolitionist from Brooklyn Law School, for example, has hired Espy to determine the races of every criminal killed in New York's three electric chairs, as well as the races of their victims. "The assignment, simple on the face, has already taken the better part of two weeks," says Espy. An explanation of his difficulties leads to a long digression on William Kemmler—the first person to die by electrocution, in Auburn, New York, on August 6, 1890. I learn much about the man that is nowhere in the vast literature on his case: from the alias he lived under to the little-known fact that his victim, almost universally misidentified as his wife (whom he had abandoned), was in fact a mistress. "His story," says Espy, "was dramatic, to say the least."

"Why?" I ask. "Wasn't he just a drunk who killed someone with an ax in a stupor?"

"A hatchet," he corrects. "And, no, I don't believe he was *just* a drunk. Murder, of the state-sanctioned type I fight, or the crimes that it is meant to fight, is never *just* anything. It is an unspeakable drama. Its impact is of the religious variety."

Espy's *Confirmed Innocents* file is thinner than the list given in the authoritative 1987 study by professors Hugo Adam Bedau and Michael L. Radelet. "Theirs," Espy says, "includes Bruno Hauptmann, executed for the death of the Lindbergh baby, which I'm afraid I can't concur with. They were right about Sacco, though perhaps not Vanzetti. But in their case, confirmed innocence isn't the issue so much as the conduct of the trial judge, Webster Thayer, who announced to the courtroom, 'I'm gonna burn those anarchist sons of bitches.' Bartolomeo Vanzetti may have been an anarchist, but he was not a son of a bitch. He was a beautiful man. Webster Thayer was the son of a bitch. You've come to ask me about executioners. Sometimes you really don't have to look further than the trial judge. Now do you?"

Espy's *Executioners* file is also very thin. "They're a rare breed. There've been dearths in this country, where the lesser of two criminals hanged the other, or where a condemned escaped death by becoming the hangman. That scenario is typical throughout history." The file's contents are mostly clippings from pulp detective magazines of the first half of the century, one of Espy's major tools for researching executions. The house is stacked with them: *Official Detective. Best Detective. True Detective. Fast Detective. Best True Fast Detective.* "These articles are of uneven reliability, but they're all I have to offer," Espy says as he hands me the file. "You're headed into terra incognita. A lot of stuff is unknown and a lot of what's 'known' is just plumb made up."

"I Executed Miles Fuller," the oldest clip, is the anonymous account of how an out-of-work steward from the Montana Cooks and Waiters Union became a hangman. It provides "no facts otherwise unknown" (at least to Espy), and "none of the real flavor of a nineteenth-century hanging," but it introduces a first, simple motive for wanting to execute: curiosity.

> Other men were waiting there [in the Union headquarters] . . . smoking, reading, playing cards and talking. The topic of conversation, for the most part,

was the approaching execution of . . . Miles Fuller. Rumor had it that Sheriff O'Rourke needed an executioner: a man to cut the rope at the hanging.

No one there seemed to want the job. . . . One man said he wouldn't do it for all the money in the world [and] that if he did, he knew he would never get another night's sleep. I laughed at that. And then, suddenly, it put a thought in my mind.

How would it feel actually to kill a man?

"Electric Death Wears a Mask," from the January 1928 *Startling Detective*, an unintelligible meditation on electrocutioners by "author-historian and award-winning academic" Curt Norris, conveys the prurient odium and garishness that have always surrounded the perception of executioners. "My good friends within these ivy walls are sheltered men and women, wise in theory and isolation, and far removed from the horrors of a blood-soaked room and the dismembered aberration of what was once a human being." A more common shade of purple is found in "Man of Doom," an account of Rich Owen, a "hard-as-nails" prison screw and former boxer, miner, and electrician who worked his way up to sergeant at the Oklahoma State Prison at McAlester. Owen designed the electric chair in the prison's basement death chamber, then took charge of the death house in 1917, when the state's traveling executioner showed up too drunk to pull the switch. He kept the job until 1948, when a "malignant disease began to burn inside him, gnawing away his vital organs. . . . But even on his deathbed the old guard's jaw still thrust forward pugnaciously. He died planning to leave his bed and perform the execution of Lewis Grayson, a Negro." The article offers a second clue as to why anyone would want to kill for the state: Most modern executioners were hobbyists, men who came to the work largely because of their interest in machines, physiology, electrical current, contemporary plumbing, etc.

For example: "I Help Them Die," from the April 1937 issue of

Front Page Detective, an as-told-to confession of George Philip Hanna, "Humanitarian Hangman" from Epworth, Illinois. Hanna was obsessed with nooses, gallows, hoods, and restraints as a teenager. He spent his days binding, masking, and hanging dummies and sandbags from the hayloft in the family's barn, teaching himself the proper ratios of body weight to rope length: Too short a drop led to strangulation; too long, to decapitation. Hanna was "unexpectedly given the opportunity to show the world what [he] had learned" when, at age eighteen, he attended a hanging in a nearby county. Horrified by the sheriff's bungling, he made his way through the crowd and walked to the foot of the gallows.

"Could I lend you a hand, sir?" he asked. "You're doing this the wrong way."

His perfect hanging that day became legend among sheriffs and wardens across the nation, and he carried on as a traveling hangman and execution adviser for forty years, touring the country with his ropes, handcuffs, and hoods, as well as a collection of weapons used by murderers he'd hanged—the only compensation Hanna asked. He was a gentleman farmer, well off by all accounts. One story has him living in a hunting cabin in a northern Illinois virgin forest; another, in the January 1933 *Master Detective*, has him owning "2500 acres of the finest land in Southern Illinois."

A jaundiced description of the executioner's house, I notice, is always offered. The man in question may have been goonish, like Frank "The Human Butcher" Johnston, a miserable hangman who was eventually executed for horse thieving, or highly esteemed, like Robert Greene Elliot, "America's Executioner," whom we'll meet below—but the dwelling must be dark, obscure, the locus of a pariah, where strange things are said and done. In "The Macabre Career of Henry the Hangman," a cub reporter from the New Orleans *Item* tracks down the squalid quarters of state hangman Henry Meyer: "two rooms behind a vacant store . . . sparsely furnished [with] an oil lamp, a table, a bench, and a walnut double bed. . . . Meyer was lonely [and] did not want the reporter to go. He said: 'I had rather ride a street car than anything.' A neighbor accosts the reporter after he leaves: 'For Christ's sake, don't put it in the paper. People around here don't know he is the hangman. Everybody will move out if you put it in the paper.'"

Also invariably offered is the man's compensation. If the capital crime involved a theft, embezzlement, or some other scheme for money, the sum is compared to the executioner's fee—the implication, clearly, that both the condemned and his executioner are murderers, looking for a payday. Most European nations had private, hereditary headsmen, but American hangmen tended to be law enforcement professionals who by force became amateur hangmen—future president Grover Cleveland hanged two men in the early 1870s, as sheriff of Erie County, New York. The early electrocutioners and gas chamber operators, on the other hand, were civilian experts. Few states had their own, and traveling executioners were the norm.

"That paucity, mind you, is not for want of zeal," Espy says. He shows me a column filler in the May 1942 issue of *Special Detective*, regarding the thousands who applied for the job at Sing Sing after Robert Elliot died in 1940. This, despite the fact that Elliot's house had been firebombed in his second year on the job, and that the previous executioner, "a mild little man named John Hulbert," had blown his brains out two years after retiring. Twenty percent of the applicants, the item notes, were women, many claiming electrical expertise; one wrote that she hated men, and "wouldn't mind bumping off a few in the electricity chair."

"This little item," Espy says, "shows how difficult it is to keep the facts straight." Elliot's predecessor was in fact named Hurlbert, his suicide came three years after his retirement, and he was hardly mild. Over the course of 140-plus electrocutions, which he performed in New York, New Jersey, and Massachusetts, Hurlbert became increasingly rancorous, perfectionist, and secretive—"The Man Who Walks Alone," the newspapers called him. There is no known photograph of him. His suicide, performed in the basement of his Auburn, New York, house with a .38 revolver, owed more to the recent death of his wife than to any stress brought on by his job.

He was not, according to his successor Elliot, a particularly deft executioner. Elliot faulted Hurlbert's choice of voltages, which led to burning and "bodies hurtling out of the chair. . . . I've never busted a strap," Elliot boasted. "My method is a rhythmic plan. I work the switch in and out and the current flows steadily, but quickly enough to paralyze the heart and brain instantly."

Elliot was a tall man with a long, Mannerist face and a large, eclectic wardrobe of bow ties, heavy wools, and wide-brimmed hats. In his dress he took after his mentor Edwin F. Davis, who designed the first electric chair, pulled the switch on William Kemmler, and trained both Elliot and Hurlbert. Born in a small New York farm town in 1874, Elliot was fascinated from childhood with electricity. From age sixteen, when Kemmler's execution dominated the newspapers, he had a ravenous curiosity about the electric chair. He wanted to know how it would feel to throw such a switch.

After graduating from school, he applied for a job as Davis's assistant at the Clinton Prison in Dannemora. Hired as an electrical apprentice, he soon became a death house protégé as well, assisting in some of the 250-plus executions Davis carried out before retiring. When the executioner's job went to Hurlbert in 1914, Elliot retired from Dannemora and for twelve years took on electrical contract work. He got the job after Hurlbert, and within four years had become executioner for the Northeast (except Connecticut—the state still executed by hanging, which Elliot considered barbarous).

Like his famously humanitarian boss at Sing Sing, Warden Lewis Lawes, Elliot did not believe in the death penalty. He eventually assisted in killing or himself killed more than 500 men and women, but he felt executions were pointless. It pained him to see prisoners as they entered his various chambers, "eyes raw and blinking under the harsh lights" after months on the row. "[T]he swaggering tough guys whose jaunty poolroom bravado crumpled, who babbled and collapsed at the long 'last mile' . . . the white-lipped frightened kids who took the wrong turns of the road and died with a half-finished prayer on their lips . . . and the women. Five of them. . . . It doesn't do any good. It never will." He felt strongly, however, that the chair, particularly if run by an expert such as himself, was the way to do it. His motives in killing, if born of curiosity, were at least part altruism and part vanity. In both appearance and tone, he can be likened to another unwilling American mass killer, Robert J. Oppenheimer, saying, "I am become Death," as he watched the first atomic bomb test.

In his second year on the job, Elliot's altruism was tested and his vanity enhanced when he traveled to Boston to execute Sacco and Vanzetti at midnight, August 23, 1927. Slipping out of the Charlestown prison a few hours later, he passed unnoticed through 800 policemen, the Boston fire department, and several thousand demonstrators waiting to follow the bodies to the funeral home for cremation. Though Elliot's sense of personal responsibility and power must have been enormous, he felt his was a guiltless hand. He had performed a specific, technical action by which the public will, right or wrong, had been enacted. "You have done it through the laws you've passed," he wrote in a memoir completed shortly before his death. "Judges and juries, people who have represented you." Threatened countless times through hate mail, Elliot wrote that his only mortal terror was of having to sit one day on a jury that would bring in a death verdict.

"I know you've come down south to learn about executioners," says Espy. "We had our traveling executioners, and they were a colorful lot. But historically, the long-standing executioners seemed to be up north. The national press paints us as being such cold-blooded killers, but it's just not accurate. A warden in North Carolina in the mid-thirties, I believe, committed suicide after electrocuting two black men. He couldn't deal with it. And a warden in Tennessee in the teens and twenties used to come out to address the crowd outside his prison on hanging days and make a big speech against the penalty. In Georgia, the governor took to commuting sentence so regularly they had to take that power out of his hands. I make the point not only out of pride as a southerner, but because large differences in the way executions are done—and they vary *so* much from region to region—are in themselvs a strong argument against the death penalty."

"Why?"

"Because it has to be *equally applied*. That was one of the challenges with *Furman*, the arbitrariness of the penalty, and part of the resolution with *Gregg*, though I don't believe *Gregg* offered any resolution at all, just a green light. But why do some states still fingerprint the deceased after his execution? And why do some give him a physical before he enters the death chamber? And fine, if you want to use diapers or a penile catheter or a rectal plug to stop a man soiling your gurney or your chair, I guess that's your business. But why do some states remove the catheter entirely while others just snap off the portion extending from the penis? Do you see what I'm getting at?"

"Not exactly."

"Well, we began by talking about Camus, the obligation to show the man's hand as he kills. You've got seventy, eighty percent of the population in favor of a penalty that to this day is carried out in the dead of night. Think about it. Intense public support for what is in essence a clandestine act. Why do they do it at that hour? And why cover their faces with their little veils and do it in subterranean chambers? Is it just an anachronism? I think they're a little ashamed of themselves. And I think it's just a bit fetishistic. In the Southwest, Arizona, New Mexico, Nevada, for example, there's a marked and very disturbing trend for young witnesses of executions to kill themselves just after highly publicized executions. I don't know how many cases I have where men take their teenage boys to a hanging and within two weeks these boys had either hanged themselves or some little friend. I have another clipping of a fifteen-year-old in Arizona who heard a vivid account of an asphyxiation, went out, slaughtered the first two migrant workers he could round up, then announced, 'I want to go to the gas chamber.' Being fifteen, of course, he didn't get his wish. And you can talk about deterrence all you want, but the studies show that homicides increase after publicized executions—and it doesn't matter if the condemned was popular, like Caryl Chessman, or a villain, like Ted Bundy."

He takes a pull of water. "Think about it. Eighty percent of the population in favor of something that may just be a fetish. You're talking about a deep rift in the national psyche. Because it shows there's something buried, like all fetishes, a memory or a wish or whatever it is that's too painful for the mind to deal with." He smiles. "But of course that's just my abolitionist opinion."

The Death Belt bears out Watt Espy's contention that "society actually kills best and most frequently when it knows most exactly what it's saying and doing." It is probably no coincidence that the states that kill most frequently are the most open about it. Though the last public hanging occurred in Kentucky (1939), Louisiana and Mississippi were the last states to institutionalize executions behind the closed walls of centralized prisons, and continued to conduct electrocutions that were open to the public, with some discretion, until the 1950s. Post-*Gregg*, the executioner's name is public record only in Mississippi, where it has been on file at the governor's office since 1938, and Alabama, where the Atmore warden has pulled the switch of the electric chair since it replaced the gallows in the 1920s.

Louisiana's twentieth-century traveling executioners, Henry "the Hangman" Meyer and the electrocutioner Gradys Jarrett, were well known to the public. The state's post-*Gregg* electrocutioner, who executed twenty men between 1983 and 1990, did work under a pseudonym, Sam Jones (the name of the governor who presided over the state's shift from hanging to electricity in 1939), but he became very much a public figure, and proud of it—so much so that he was fired when Louisiana switched to lethal injection in 1990. If Sam Jones knew he was to be introduced to you at a party or picnic as the state executioner, he'd make his way unseen to the beer bucket, put his right hand in the icy water until it was cold, wipe it dry, then give a grim, hollow stare as he shook your hand.

Texas, by far the capital punishment leader in the post-*Gregg* era, does hide its executioner's identity, as does Florida, until recently the second-most-lethal state. Texas, however, makes such show of its protocol and techniques that the executioner seems, as the saying goes, to be hiding in plain sight. On my first visit to Huntsville, I was told, on pretty good authority, that two men eating chicken-fried steak next to me in a downtown café had taken part in a lethal injection the night before. Death row has a Media Day, reporters are freely given tours of the death house, and each execution comes with regularly updated fact sheets, similar to the handouts in sports-events press rooms. They detail oddities, stats, and facts of the man's crime, his stay on death row, last meal, last words, and his execution.

Florida, on the other hand, takes the executioner's secrecy to such extremes it seems like a red herring. One of the last states to use a civilian executioner, the Sunshine State is also one of the last that hoods the man. Hired through classified ads, his name is known by only two people in the state, whose identities are also secret. At 5 A.M. on the morning of sentence, the executioner is picked up, hooded, at a designated spot by an administrative assistant of the Department of Corrections (DOC). The hood stays on for the drive to the prison farm at Starke, where the executioner is shown to a small room off the death chamber. He sits there until sunrise, when he's summoned to another small room called the "executioner's alcove," which is visible to the execution participants but not the witnesses.

"It is a strange sight indeed, that man sitting there in his hood," admits the Florida DOC's unusually jocular spokesman, Eugene Morris, whom I speak to several times in an attempt to interview the executioner. "Particularly in this day and age. Not to mention at six in the morning. It leaves you with no doubt, though: today's going to be different than yesterday."

I tell Morris I really want to meet this man. "No, sir," he says. "Not on this side of life."

"Can you tell me his name?"

"Sure I can," he laughs. "But then I'd have to kill you."

"Is it the same man each time?"

"Don't even bother asking."

"Does he look like something out of the past, in his hood?"

"Yes. I would say so. Very much."

"Does it seem a little fetishistic?"

He laughs again. "I'll have to get back to you on that one."

How strange this hooded man must look to the condemned, who in Florida has full view of his executioner before his head is pinioned to the chair, is captured in David von Drehle's account of John Spenkelink's final moments in *Among the Lowest of the Dead*:

> In one swift motion Spenkelink was thrust into the chair and the practiced hands began moving over him, cinching the leather straps tight. "We came in and we strapped him into the chair and I remember John was looking all around the room, almost like he was curious," [Warden] Brierton recalled. "He turned his head and looked back at the executioner standing there [in the executioner's alcove] in his black hood. He just stared at the guy. And then, I believe, his nerves started to go."

After the condemned is strapped in, two electricians engage the circuits and a third man throws a switch activating the "executioner control panel." When the moment comes, the warden nods to the hooded man, who hits a switch that begins an automated sequence of voltages. For reasons no one in the DOC can explain, this hooded man is the last to leave after sentence is carried out. Driven to the spot at which he was picked up, he's paid $150 in cash.

Why such pains to hide a man hired through a classified in a fifty-cent newspaper? A man who does nothing but turn a switch when he's told? And if his hood is a red herring, what is it deflecting attention from?

The rather simple answer—it is deflecting the state's attention from the fact that it is killing, precisely at the moment that it does— emerges from a series of mind-numbing, double-speak conversations with prison/DA spokesmen of post-*Gregg* execution states. Among the regional quirks and anomalies, I notice a tendency to identify the executioner, when pressed, as "the DOC itself," a bureaucratization that not only veils the switch-puller's identity but makes the state appear less vulnerable to the taint of being unworkably rigid, unduly harsh, cruel and unusual, or arbitrary. Bureaucracies *are* harsh, rigid, arbitrary, cruel, unusual. Why should the bureaucracy of the death penalty be any different? In fact, it is vastly more so. The multipage protocols—from the unit prices of thiopental "kits" to the exact times and "sally ports" through which the executioner, witnesses, and condemned arrive and depart—are surreal in their specificity, filled with obfuscatory details that in the long run are probably little more than the state's self-absolution.

This bureaucratic distancing is seen in other post-*Gregg* trends: putting physical distance between death row and the death house; dispersing responsibility among an "execution team," ranging from Georgia's three men to the State of Kentucky's fifteen. Oklahoma uses three lethal injectioners, each of whom injects one of the drugs. Several others, Georgia, Missouri, Utah, and Illinois among them, use the old standby of the blank bullet placebo: three lethal injectioners, only one of whom has the lethal drugs in the drip bags on his IV stand. Georgia, which still offers the electric chair, also has three keys that activate the switch to it.

Tennessee's electric chair, activated by one of two men turning keys simultaneously in a blue enamel box labeled Electric Chair Control, has computer software that randomly decides which of the two keys starts the voltage cycle. It's a nice sci-fi touch, meant to hide the executioner's identity from himself, but every sci-fi reader sees through the ruse: computers, by definition, cannot create randomness. Therefore, the author of the software is the executioner. Tennessee's answer for that is simple: the author of the software doesn't know what he wrote it for. New Jersey's post-*Gregg* death penalty statute has a measure, written by a Metuchen dentist who served as an assemblyman, that mandates that "the procedures and equipment [of a lethal injection] shall be designed to ensure that the identity of the person actually inflicting the lethal substance is unknown even to the person himself." In Texas, he is no longer even called the executioner; he is the "designee of the Director."

Even if you are one of fifteen anonymous Kentuckians who accept responsibility of a state-sanctioned murder, however, the question remains: Why be in that position at all? One of the oldest, most common answers was money. Robert Elliot was a perfectionist humanitarian who took great pride in his work and felt he served the condemned, but he also made a lot of money: $50,000 in less than fifteen years, enough to buy a house in a wealthy part of Queens, New York, in the middle of the Depression, a property with enough land to breed a large variety of roses, Elliot's great passion in life. Thus, the third post-*Gregg* exculpatory tradition: no profit motive. An increasing number of American executioners are volunteers. They don't even get overtime. In Texas, that adds up to several hundred "pro bono" hours a year for each man on the team.

When witnesses to Huntsville's lethal injections try to look into the plate glass window behind which America's most prolific executioner stands, they see themselves: it's a one-way mirror. The metaphorical suggestion—we are all the executioner—was unintended, says James A. Collins, executive director of the Texas Department of Criminal Justice, but it is nonetheless apt. Many people still want to pull the switch. Georgia advertised publicly for executioners in 1985, as did New Jersey in 1991, and both were besieged with applications.

The Huntsville death house stands behind a red-brick retaining wall on the corner of Avenue I and Eighteenth Street. It's not Main Street, but it's a busy corner during the day. I stood there one midnight in early April 1995, thinking about Noble Mays, the forty-one-year-old murderer I had corresponded with. Mays had been issued a death warrant for that hour, his fifteenth warrant in as many years on the row, but his lawyers had told me that he would almost certainly not go to his death and would be available to talk on Media Day.

I was looking forward to meeting Mays, an oil field roughneck who, in 1988, offered to drop his appeals and go to his death for $10,000, which his wife would get—a modest proposal that the Department of Criminal Justice turned down. A year later, he came close to becoming the first man to escape Huntsville's death row. He and his cellmate greased themselves from head to toe with hair tonic, squeezed through a ventilation duct, and made it out into a utility passageway. Mays got stuck while trying to squeeze through a one-square-foot airshaft at the end of the passageway, at which point he and his cellmate started laughing hysterically and were caught.

It was an odd feeling to stand on that corner at midnight, at the place and time of a scheduled death. "All of us must die," said Christ, "but none knows the day or the hour." It was an even odder feeling the next morning to read in the Huntsville *Item* that Mays's execution had taken place as scheduled. I hadn't kept up with his case while on the road to Huntsville. Mays's appeal had taken an unexpected turn a few days earlier, and the Supreme Court had denied him shortly after 1 A.M. the night before. He was pronounced dead at 1:42 A.M.

I visited the death house eleven hours later. Neil Hodges, the assistant warden in charge of the place, gave me my tour, his second of the day. He's a man unlikely to be found outside Texas: six-five, 240 pounds, all muscle, immaculately groomed and self-possessed, with alligator boots, tight jeans, a short-sleeved white shirt, expensive orange aviator glasses, a white Stetson, and a rather high-pitched, laconic voice. Roses bloomed in the death house's Spanish-style courtyard, separated from Avenue I and Eighteenth Street only by the retaining wall I'd stood in front of the night before. "Nice old vines. Captain Byrd's," said Hodges, referring to

Huntsville's old electrocutioner, Captain Joe "Bull o' the Woods" Byrd, an assistant warden who conducted over 200 electrocutions and had a great passion for roses. Red-winged blackbirds sat on the spreading branches of a low-slung Japanese maple, which Byrd had also planted four decades ago. It was a beautiful, peaceful place.

"We have a pool of officers from which we select our execution team," Hodges explained softly, clearly proud of the quiet decorum of the death house. "They are all calm, respectful, businesslike, large." We entered the death house and paced past the disused former death row and the last night cell to the death chamber. "They are all guaranteed anonymity, though to tell the truth, I've known them all for years and I don't think they care one way or another about it."

Stepping into the blue room, I asked Hodges if he was present the night before.

"I was," he said.

I asked where Mays was lying and he guided me to the spot, a few feet from the executioner's window at the rear of the room. The death chamber still had the Trumbull Electric 440-watt generator and step-up transformer for the electric chair. Behind the one-way mirror, the executioner's chamber had the old rheostat handle Captain Byrd had thrown for twenty years. I told Warden Hodges how I'd stood outside the retaining wall the night before, oblivious to what was happening in here, and he gave me a "Wish I'd known you were coming" wave of his hands. I told him it was difficult to believe a man had been killed right where I was standing, and without a trace of irony, he told me: "You'd have felt the same way if you'd have been in here eleven hours ago, watching it happen. I've seen quite a few, and it still doesn't look like death to me."

Though members of Huntsville's team receive no money for their work, a free-world doctor is paid to pronounce death. He has nothing to do with mixing the chemicals, loading the IV stands with drip bags, or establishing needles, tubes, and catheters. In fact, his presence is not necessary. After a 1984 AMA directive banning member physicians from lethal injections (argued to be a violation of the Hippocratic Oath's clause to "above all do no harm"), Huntsville lost temporary access to doctors. It was a minor snag. At the March 31, 1984, execution of Ronald Clark O'Bryan, the "Candy Man" (condemned for insurance-murdering his eight-year-old son by lacing his Halloween candy with cyanide), the free-world attendant was a justice of the peace.

"Our 'executioner,' so to speak, is truly the Texas Department of Criminal Justice, at least as it is represented by a team [of four men]," says Executive Director Collins, who has appointed and overseen this team for twelve years. "They all come forward, on their own, and tell me, 'I can alleviate this chore for you.' We speak regularly, plus we run through the protocol on a weekly basis. In a dozen years, I haven't heard of a single problem. Some retire. Some move on. We've had people say, 'I've had enough,' but there are always others ready to step up."

"Why?" I ask.

"I believe they're motivated by a sense of duty, to be of service to the state. I knew Captain Byrd. He said electrocutions were not a pretty sight, that he found them distressful. One of his problems, we found, was that he ran the show alone. We're a team in a true sense. And I, for one, believe the anonymity not only helps but is crucial."

Warden Morris Jones, who oversees Huntsville executions, refuses to speak to the media, an effective if unusual way of keeping his own anonymity. Even more unusual, he doesn't speak during executions. He signals the executioner to begin the flow of thiopental by removing or lowering his glasses, his face showing "the serene expression of a Christian holding four aces," in the words of a local stringer paraphrasing Mark Twain. On the morning of sentence, the condemned is brought to Huntsville from the death row in Ellis Unit. There Warden Jones, meeting him for the first time, helps see to the disposition of any property or last will and testament, and takes the final meal request. The state allows any menu within limits—"Pheasant under glass would be pushing it," says Director Collins—and offers no sedatives or even a last cigarette. Smoking is forbidden in the state penal system, including the death house.

In 1991, John Whitley, then the warden of the Louisiana State Prison at Angola, traveled to Huntsville when Louisiana decided to switch from the electric chair to lethal injection. Though the visit was highly technical, Whitley, who is the only warden to conduct both an electrocution (three) and a lethal injection (one), was most impressed by the way Texas handled the personal aspects of the execution. "It began with the condemned," he says. "Where you kept him, how you moved him, how he made out his will, the last meal request, his last words. It ended with the witnesses, because, like it or not, you are putting on a show. What we learned from Texas about that was simple: you got nothing to hide, so put him right in front of the witnesses, facing sideways. If he turns to the gallery, they get to watch him go to sleep. If he doesn't, then they don't get to watch him go to sleep."

I ask Whitley which of the two he preferred, and he responded in the same way as a number of Death Belt executioners. "I prefer a simple hanging, on a tree or a gallows right in the middle of town, at high noon. It's a good, quick, violent death, and if you do it as soon as possible after the capital crime, the message is clear: If you do that, you'll get this. End of sentence. Pardon the pun."

APRIL 18, 1995. ANGOLA STATE PRISON, LOUISIANA. Warden Burl Cain is thrilled that Antonio James is eating so heartily: fried Louisiana oysters, hush puppies, french fries, fried shrimp, crawfish étouffée, gumbo, pecan pie. The étouffée, a big potful brought in from outside the prison, is particularly good. After a mild winter, the "Mississippi mudbugs" are big and juicy this spring. So are the Louisiana oysters.

Cain is a short man with a permanently sunburned face, heavy black-frame bifocals with thick lenses, and short-cropped hair the color of ballpark mustard. In six hours, he will lead James to a chamber at the end of the long reception room in Camp F where they sit, then assistant wardens will strap James on a table, veins will be opened in both his arms, and Cain will nod to a man standing behind the windowed door of a room behind the chamber and say, "Do it."

For now, James is asking Cain if he was planning on having

another piece of pie. "I could eat one more," James tells the warden. "But only if you join me."

Cain is stuffed after the big meal, but agrees to a second piece regardless. Faith is the only way to get through something like this, he believes, and a healthy appetite shows a soul at peace. It's the Monday after Easter, and both men feel that's a good sign. The refrain of "We Shall Gather at the River" comes from a waiting room beyond the reception area, where James's family has been sitting all day, talking and singing and crying. Cain looks down at his second piece of pie and sticks a fork in it. "Antonio," he says. "Me and you's eating like hogs."

James is a heavyset man just under six feet with an awkwardly groomed Afro, cappuccino-colored skin, and a friendly, country-simple manner, though he's actually city-raised. After thirteen years on death row, he has a big, nervous smile that never really leaves his face when he's around people and a warbling voice that makes you think of Aaron Neville singing George Jones's song "Grand Tour." He hadn't wanted this meal, but Warden Cain insisted. "Man, you from N'Orleans," he'd said. "Why don't you have you a few oysters and some catfish fried up, some seafood-chicken gumbo, a little strawberry shortcake." James agreed, provided the gumbo had no chicken in it, that they'd skip the catfish, that dessert was pecan pie—homemade—and that they would eat together. A death row sergeant who brought James here is also at table, as are two assistant wardens from the execution team and a clerk from the Loyola Death Penalty Resource Center in New Orleans, which has represented James's appeals. The other three wardens on the execution team said they didn't want to eat with a man they were about to kill.

Cain, who said the grace, had no such reservations. He's never executed a man, and hopes he'll never have to, but it helps to know he won't be sending this man to hell. James is condemned, but he has been saved, and for weeks, Cain has done all he can to make this night go easy. He gave James's lawyers complete access to his prison for their last-minute appeals, sent cars down to New Orleans for his family, and talked to James regularly about salvation in

Jesus. That was like preaching to the choir, maybe even to the priest. Cain had brought his fifteen-year-old son to visit on death row, and James had given the boy a magnificent sermon on following the true path through the forest of life.

Cain and James have been talking about Christ since they were driven the five miles from death row to Camp F today at noon, then throughout the long afternoon in the last night cell, just off the death chamber. The last night cell is a terrible little place—nothing but a toilet, a cheap, quarter-inch foam mattress on a tiny metal bed bolted to the wall, with a roll of toilet paper at the foot. They talked about redemption all day and didn't seem to notice where they were.

For now, Cain wishes James's family in the next room would stop crying. After dinner, he'll go out and ask them to stop. "It's Antonio that's going to die," he'll tell them. "You really ain't making things easier." James's faith, so simple and strong, is actually somewhat unnerving. He has the quality of the holy fool, with his unswerving devotion and simplicity, and his face actually glows when he smiles. At various times during his thirteen-year stay on death row, James's IQ has measured between 65 and 75. A former junkie, James may or may not have been the triggerman in robberies during which two men were killed sixteen years ago, but today he's a sober, responsible, and genuinely nice man. "If you weren't on death row," Cain had told him at the pardon board hearing on Good Friday, "I'd make you a class-A trustee, have you working in the warden's mansion instead of wasting away on the row."

Instead, they sit at this long table in the reception room, waiting for midnight. Normally, Camp F is a relatively free-ranging barracks for trustees—long-term prisoners in good standing, and with special privileges. Set against a copse of old pine and sugar maple, a cattle-grazing pasture, and a staggered winter-crop field, it has a lazy, deep-country feel to it. The only nearby buildings, a slaughterhouse and a vegetable cannery, were closed long ago by the FDA. The reception room usually serves as a prison visiting room, as well as a store for the inmates' arts and crafts. By day, the sun comes streaming in here, making the model cars, balsa-wood airplanes, tomahawks, and headdresses on the windowsills and tables look old and faded. All the objects show a tremendous attention to detail: the different colors of each spoke of the model cars' tires, the hand-carved scrolls of the tomahawk grips, the tiering of the feathers glued into the headdresses. The people who put these things together had a lot of time on their hands. Life here feels like one long, hot summer afternoon, with nothing to do. Longer than that, really: when you get life in Louisiana, they mean it. The state doesn't even have LWOP, a life-without-parole sentence. All life sentences come without parole.

When you get death, they usually mean it too. Cain began to feel that strongly on the drive with James to Camp F—going by van past the several dozen houses, shops, and facilities for guards and officers that constitute the prison's "free-man town," where toddlers were jumping on a trampoline and older kids shagged softballs on the largest of the prison's three diamonds. Angola, 18,000 acres of former plantation land, has its own area code, radio station, and bus line to the schools in nearby St. Francisville. James, looking out the window at a world he hadn't seen for thirteen years, was particularly interested in a ridge carved out of a hillside just past the free-man town. Cain explained that an inmate pushing dirt with a bulldozer a few years back had turned over a footlong tooth in that ridge. Archaeologists from Louisiana State University came in and dug up a well-preserved mastodon fossil out of the red clay—*loess*, they call it down here. James had heard about it on death row, but of course had never seen it. He told Cain he thought he could make out the mastodon's shape in the excavation's curves. Cain didn't see the shape, but nodded anyway. He figured it might well be in there. "The beast had been in the loess over two million years," he told James.

Tonight, the minutes are flying by. James has lost all his appeals—with the state supreme court, the Fifth Circuit Court of Appeals down in New Orleans, the U.S. Supreme Court, and the judge who signed his death warrant in 1982. The pardon board turned him down on Good Friday, and his lawyers have had no luck with

Governor Edwin Edwards. At four this afternoon, the reception room was cleared of everything but the long table at which they sit, a soda machine, and a 1950s-era candy machine with the legend "Don't Go 'Round Hungry!" Mosquitoes dot the lime-green walls, just above swatting level, and shafts of the bright security lights outside slant in through the high, louvered windows. Cain can feel the huge weight of justice forcing this man to his death. It's been growing for weeks, and now it's palpable.

Alvin-Adams, the fifty-three-year-old son of one of the two men James was convicted in 1979 of killing, is in the other waiting room. "A retired, fully disabled sailor," as he identifies himself, he wears a white three-piece suit over a brown shirt and lemon-colored tie. Every time Cain goes in to see how he's doing, Adams clamors for justice, saying he doesn't believe in Antonio James's redemption any more than he believes in Santa Claus, and that he doesn't want to hear about the evils of society either, because society didn't make Antonio James shoot drugs into his arm and gun his father down in cold blood. He says society kills unborn babies but won't kill a rabid dog like Antonio James. In a histrionic gesture, he's brought $800 in twenties and tens—in case the state needs "cash money" for the execution. He's also brought his checkbook, which he's taken out and shown Cain. "If you need more than the eight hundred, I'm good for it. If you doubt my credit, I'll sit in that man's cell on death row till the check clears," he said. "Just let me watch the man die. If you hold it in hell instead of here, I'll be down to watch."

At seven-thirty, James asks Cain for help drafting his final statement, which he will read to the assembled press in this room four hours later. Now that dinner's over, they'll soon be setting up a microphone stand in the middle of the room, where their table is.

Cain is honored that James has asked for his help. They write about death, redemption, and salvation, and both are able to quote freely from scripture. "It's like we're ministering each other through this," Cain says quietly. "Nothing to be afraid of, is there?"

"I am a little afraid," James admits. "Mostly, I guess I'm afraid of my fear of death." He says he thinks that after a period of shock he'll see God, but he's not sure exactly how.

Cain shakes his head, unsure what James means by *shock*, then thinks hard before explaining the execution. The shock that he thinks James is speaking of was with the electric chair, which he's glad they don't use here any longer. Cain had witnessed an electrocution in Camp F's death chamber a decade ago, when he was warden of a medium-security facility in the southern part of the bayou country, and he's never forgotten it. The man had a red bandanna in his right hand, and Cain remembers how the hand seized on it with the first jolt, how the bandanna shot out of his fist after the second, then how the hand opened lifelessly with the third jolt and what looked like a steamy river flowed down his legs as his blood boiled. The sweat, if that's what you called it, seemed to tint the man's jeans red, like it was mixed with blood, and made Cain think of the slaughterhouse in his hometown. As a child, he was always terrified when he'd watch the butchers put electricity through the carcasses, hanging headless or in quarters, to tenderize the beef. The beef slabs jumped like they were still alive, just like that man did. Cain didn't know how Jesus felt about him being in that room when the man got killed, over and over, and he has to admit he really doesn't know how Jesus feels about him being here now. In fact, Cain has no idea what he's doing here. All he knows is that he has to be strong for Antonio James.

"After I tell that man to put the poison in you," he finally says, "I'm going to say, 'Okay, Antonio, He's waiting for you. Now close your eyes.' The face of Jesus will come before you, and when the poison hits your heart, you won't feel shock or pain at all."

James doesn't say anything. Cain thinks about the gurney. It's a little over six feet long, with thin, padded rests for the man's arms, sticking out downward at forty-five-degree angles three-quarters of the way up, and a raised oval on the bottom, in case the condemned begins to struggle. If he does, his feet are put over the oval, taking away any mechanical advantage in his hip muscles. The gurney is made so it can stand up. Cain thinks about how James's feet would probably rest on the oval if it was made to stand with him on it. It would look just like a crucifix.

"Antonio," he says, "just like Jesus Christ on the cross, with the thief right there by him, what you're going to see is a band of angels. They're gonna take you straight to Heaven. You ain't going to feel no pain or shock. They took care of that a long time ago."

Tonight, the miracle comes in a different form. Summoned for an emergency phone call at 8:20, Cain goes into the death chamber and picks up one of the two red phones on the wall to the right of the executioner's chamber. This one is the hotline to the district attorney's office. He asks the man at the other end for a pre-arranged codeword, to identify him beyond doubt, then hears the best news he ever got to tell a man. The Louisiana Supreme Court, reversing itself by a single vote in light of evidence that James's lawyers have found over the weekend, has granted an eleventh-hour stay. After all these years, they're still not sure that James was the gunman in the fatal holdups.

Cain goes to the two waiting rooms to tell first James's family and then Alvin Adams the news, and to escort them to prison cars that will take them back to the administration building by the prison's front gates. Back in the reception room, he slides an arm around James's shoulder. "I don't want to build your hopes up, Antonio," he says, "but you got you a temporary stay." He feels justified in his faith when he sees how impassively James receives the news. He's smiling, but it's just that shy, dopey smile. The man hasn't shed a tear or made excuses for what he did or didn't do, and Cain knows he'll have no fear for Antonio James's soul when he finally does send him to death a month or a year from now—too many people in the state want him dead for that not to happen. He's just happy he didn't have to do it tonight.

"The governor's board turned you down on Good Friday," he says as James puts his final statement in his denim shirt pocket. "But, hey, you got a free meal on me the day after Easter."

James slaps his gut, then presents his wrists for the sergeant's shackles for the five-mile drive back to death row. It's his thirteenth stay of execution since arriving on the row in 1982, but none have come this close. Thirteen stays in thirteen years, thirteen steps up the hangman's gallows, thirteen loops on a hangman's knot. Lucky thirteen. "Warden Cain," he says, slapping his gut again, "you know we can't eat like this all the time."

The day after James's eleventh-hour reprieve, I visit Wilbert Rideau and his fellow editors of the inmate magazine *The Angolite.* Their office, a long, dark room below ground level of the prison's largest unit, has three desks, a computer with a hard drive that crashes on a regular basis, a supply closet, and one barred window, through the top of which Rideau feeds a stray cat who has adopted them. Prison newsletters, typically four-page folded and stapled mimeographed sheets, usually serve as the warden's mouthpiece or for highly censored inmate complaints. *The Angolite*, a seventy-plus-page magazine given top honors by the American Society of Magazine Editors, has a half dozen candid features per issue on normally taboo subjects ranging from executions to homosexual rape. One particular exposé, of Robert Wayne Williams's botched 1983 execution, was instrumental in getting Louisiana's electric chair abolished.

All six men on the masthead have been on Angola's death row at one time or another, but as cold-blooded killers go, these editors are a strange covey: well-read, and brimming with a will to work. They're also dead serious and mindful of social nuance. In an earlier telephone conversation, Rideau, telling me how to establish credentials for this interview, went out of his way to advise me that the assistant warden I'd be speaking to has a stutter, and to mind myself. "It's a physical defect," he admonished, "not a personal failing." He also directed me to "the only bed-and-breakfast worthy of the name" near Angola, detailing the breakfast I could expect and waxing on the beauty of the live oaks overhanging the drive up to the old plantation house, where the owners live. After hanging up, I realized he had never seen the property. He landed on death row in 1961, stayed there until *Furman,* and has been in Angola's general population since. The stuttering assistant warden whose sensitivities he was protecting is a prominent member of the execution team.

Rideau, whom *Life* called "the most reformed man in America," has a strong physical and vocal resemblance to Richard Pryor, but no discernible sense of humor. I try a few jokes early in the visit, then realize that he and the other men in this office are not interested in laughter. He makes introductions formally: handshakes, even slight bows, are offered, then things like cigarettes and, later, an *Angolite* T-shirt. "Michael Glover," he says, introducing a man with a Brooklyn accent who has taken on much of the workload of late. "And Douglas Dennis," an immense, awkward man with a glass eye who comments several times on the old-fashionedness of my Pall Malls but refuses repeated offers. Two other men in the office ask not to be mentioned by name. They are interviewed often, they later tell me, and don't care for the condescension usually shown them by reporters, whom they look upon as colleagues with better jobs but none of the fame, awards, and exclusives *The Angolite* has achieved.

They also don't seem to want to discuss executioners. Rideau and Dennis had seen the old traveling executioner, Gradys Jarrett, on death row in the 1960s, after Louisiana took its traveling electric chair off the road and put it in the Angola death house. A pale Texan "with the coldest damned blue eyes you ever saw," as a former warden described him, Jarrett had held the job since 1940, when the state switched from hanging. Though he had always been known as a highly professional and even courteous executioner (Jarrett actually said good-bye to each man as he pulled the switch: "Good-bye, Joe"), he had taken to drinking heavily after twenty-five years on the job. Before executions at Angola, he would down a half pint of whiskey, then visit the condemned's cell and ask the man what his last meal request had been. "If that had been me," he'd say each time, "I'd have ordered kangaroo tails from Australia. Well-done." He'd drink the second half of his whiskey after throwing the switch.

Rideau also doesn't want to discuss the man who replaced Jarrett when the penalty returned to Louisiana in 1979, the Baton Rouge electrician who worked under what they call the "nom de chair" of Sam Jones. They even know the name of Angola's lethal injectioner, but won't tell me. "The man has a family," demurs Glover, whose workload is equal parts editing, layout, and reminders of office responsibilities that sound straight out of some 1940s etiquette book: "Remind _____ to thank _____ for the timely information he conveyed about _____"; "Make sure you let _____ know that his contribution, while excellent, was unfortunately duplicative of an earlier submission." Throughout the day, there are inchoate nods regarding their complex relationship to their jailer-publishers. I'm reminded of a Louisiana leper colony I'd read of— how caricaturedly socialized outcasts can be when forced to confederate for a lifetime.

They consider themselves no less unfortunate. "The common wisdom about murderers," says Rideau, "is that we are a special type, lurking in society, like a virus, like aliens. The truth is that murder is one of the great common denominators of human society. We are the only species that kills for anything besides sustenance."

"Actually, orcas kill for revenge," I say.

He shoots me a glare that makes it clear he really doesn't want to hear about orcas.

"You can even argue that we are more human than others because we *have* murdered," Glover adds. I smile when he says it, but he doesn't.

"In order to keep most of us from doing what all of us want to do," says Dennis, "some of us will forever suffer. Are you hearing what we're saying?"

I tell him, honestly, that I hadn't even thought of them as murderers until they brought it up.

"And now you're scared," Dennis says.

"Yes."

He gets up abruptly from the desk he's been leaning on and lum-

bers over. "All right," he says. "Let me have one of those Pall
Malls." I offer the whole pack. He takes three, puts one behind
each ear, lights the other, and French-inhales the smoke in a way I
haven't seen for twenty years. "The first thing to understand about
an execution," he says, "is that it's a ritual. You've heard that
before, but probably only as a truism. There's nothing cliché about
an execution. It is a modern religious ritual, sanctified in the sense that
everything represents something else. The executioner's fee is the eye
for the eye. The sacrifice is implicit in the fact that only one out of
every hundred or so gets killed. The blood atonement is what the prisoner's last meal symbolizes. Notice how the exact menu always gets in
the newspaper story? That's just some AP asshole, acting out a ritual
he doesn't even know is a ritual. He's covering a communion."

Dennis is a middle-class northerner, a black sheep who at the age
of nineteen killed a man accidentally while sowing wild oats in a
barnstorming through the South in the late 1950s. He wound up in
Angola and lost his eye in a knife fight in which he killed an inmate,
for which he received a death sentence. Like Rideau, who taught
himself to read on the row, Dennis spent the 1960s reading everything from the *New Republic* to Dostoevsky. Everyone in this office
has read Dostoevsky, in fact, and to a man they are convinced that
he murdered someone during his youth. No one, they say, could
have understood the psyche of a murderer that well without having
tasted blood himself.

Like the other men in this office, Dennis got off death row with
Furman. Within three years he had managed a partial parole, thus
becoming probably the only man ever to have worked his way out
of both a death sentence and a life sentence at Angola. He made it
to a Baton Rouge halfway house by the mid-1970s, studied computers at LSU, and became something of a cause célèbre. A local
assistant DA hounded him, however—a man Dennis says wanted to
make political capital out of a formerly condemned inmate's attending college at taxpayer expense. In 1979, a routine traffic pile-up on
a Baton Rouge bridge became complicated by the assistant DA,·
who told Dennis that the accident was all he needed to send him

back to Angola. Dennis turned fugitive and established a fake identity, complete with passport. Over the next ten years, he went to
California, married a German woman, joined Mensa, and began
working in artificial intelligence, writing software for prototypes of
the early intranets for a Palo Alto firm. He was caught in 1988 and
brought back to Angola.

As the above précis would indicate, Dennis is a baffling man, a
personification, to paraphrase Kafka, of the charisma of the condemned. He speaks with a loping slowness, and the bad prosthesis
of his glass eye makes him look frightening, but his thoughts are
crystalline, coming out in perfect declarative sentences, and his
manner is gentle and seductive.

"Religion," he says, "makes ceremony of birth, death, and rite of
passage. At Camp F, we make ceremony of the expulsion of taboo,
and in its mock privacy, we have the satisfaction of the totem. The
highest form of religion is personal embodiment, whether it's of individual faith or the anachronisms of implied pantheism, which the
death penalty is, in my opinion, our strongest link to."

"And how does the executioner figure in this ritual?" I ask.

"Different ways in different times. In most early religions, in the
terrified exaltation of men whose brains didn't exist in two parts,
God was the executioner. It's in every early faith, from Abraxas to
Zoroaster, and it's in the Bible, too. Maybe with decreasing frequency from the Old to the New Testament, but it's there, from
page one straight through to Christ on the cross and the
Consummatum est. You killed the man by raising him to the heav-
ens, on a crucifix, on the scaffold, an altar. It was human sacrifice,
and God snatched the man up. In godless times, like the Inquisition
or the Reformation, you had dungeons, and the task was hidden.
You paraded the men through a kind of pageant in the central
square in an auto-da-fé, but you killed them after nightfall, outside
city walls, and the men who did the killing wore masks and performed, and were rewarded with applause and gasps while they
tortured the condemned. Beginning with enlightenment and indus-
trialization, when executions became progressively archaic, the exe-

cutioner became a whimsical figure of creepy dread, a kind of memory. Cartoony. Halloweenish. Friday the Thirteenth. In many respects, that stayed the same until the modern period, when we bring executions up to ground level, sanctify them with the trappings of the operating room or the bureaucracy of the state, and all you have left is the symbolism."

"Symbolism of what?"

"The symbolism of impersonality. We sanctify the executioner with impersonality. We say: He is the hand of the state, and must be protected. He is retaliating, for us decent people, against a man who has left society, a man who has repudiated citizenship by committing the act of murder. Bull, fucking, shit. The executioner is not a figure of vengeance, and his hood is not disguising his identity, because everyone knows who's behind it. The hood is his sanctification, and what it sanctifies is denial. You can see that in the mockery of the nineteenth-century hangmen, when people had begun to be ashamed of hangings and made the hangman into a sacred goat and threw food at him."

"Denial of what?"

"Denial of taboo, of the last taboo left in this society."

"Which is?"

"Which is the urge to kill. Which is what the condemned has acted on, and what society acts on every time it executes. In denying that you are committing taboo, that which cannot be looked at in the face and therefore must be hooded, you deny the terror that the urge to kill brings you, and you also deny the extreme allure that anything that has been denied takes on in its closeted space. When I asked you before if you were afraid, you were honest, but you didn't realize what you were admitting. I hope you never do," he says, lighting the last of the Pall Malls. "For the sake of some random member of our species. Not to mention any killer whales you might happen upon."

"The executioner," Rideau says, "is a man who appears only when society turns its back on the urge to kill. In the 1960s, and in the afterglow of that period of freedom, he went unemployed for the first and last time. You could see the significance of his absence most clearly in the moments of change, like in the headlines after *Furman*. They were the first I'd seen in eleven years, coming off the row. It was almost like reading about a change in the evolutionary ladder.

"Now, in their benightedness, hiding from the truth about themselves, society has become a lynch mob again. They exhort their leaders to kill with the seventy-five percent pro-penalty polls, which measure nothing but collective terror. And they are exhorted in turn by politicians. The whole exchange would be impossible without the media, who flock to the gallows in a kind of feeding frenzy. We'll go six months sometimes without a word from the outside world. Then a date, like Antonio James's, will come up, and there'll be a line of you people trying to get in this office."

"So you're saying I'm the executioner?"

"That's tempting. But we know exactly who our executioners were. We saw them in the courtroom, sitting behind the desk across from us, on the right side of the court. Being on the row teaches you that capital punishment isn't anything like it's normally construed. It's not about crime, conviction, punishment. It's not that linear. It's a slingshot. It's a weapon that never goes off when you think it will, and never goes where you aimed for. It's completely random, except of course for the fact that it's you, sitting in there and waiting for it to come, and that's as unrandom as you can get.

"We interviewed Sam Jones. He said and did a lot of odd and colorful things—painting pictures of the men he pushed the button on, talking about how killing people was no different than going to the fridge and getting a beer out. But he wasn't an executioner, not in the way we've been speaking of, not what you're looking for."

Rideau gives me the name of an assistant district attorney and a phone number in St. Francisville Parish, across the river from New Orleans. "That's the man you're looking for," he says.

Jim Williams is a meticulous, ambitious prosecutor who came up through what he calls the "breeder factory" of the New Orleans

DA's office. He has prosecuted one-eighth of Angola's current death row, a figure he glories in. New Orleans, a mile across the Mississippi, has posted the highest murder rate per capita every second or third year since *Gregg* came down, but the New Orleans DA, Williams's former boss Harry Connick Sr., has secured only one death penalty conviction in those years, despite a phrase that made him famous down here: "We need the death penalty to show that life is precious."

Williams's corner office, overlooking the New Orleans skyline from the fifth floor of St. Francisville's municipal building, is adorned with execution tchotchkes and mementos. On the wall above his desk are clippings: a *Hustler* cartoon of two guards carrying a floral wreath to a man awaiting electrocution, with a banner reading "Fuck You"; an underground magazine article chronicling prison rape in detail; a clipping of a *New York Times* story about a Cantonese woman who was tried and executed the same day her husband was murdered. "That," says Williams, pointing, "is deterrence. In a nearby parish, there was a capital conviction that took only one calendar day—voir dire through penalty phase—but the guy will be sitting on death row for years regardless. That is failed deterrence."

Stays and abeyances are the bane of Williams's life. "If anything makes me mad about these abolitionist appellate lawyers," he says, waving a dismissive hand northward at the window behind him, perhaps at the offices of the Loyola Death Penalty Resource across the river, perhaps at the entire country beyond, "it's how they keep these guys alive for eleven years, average, at huge expense, then talk about how all that time on the Row constitutes cruel and unusual. The manipulation of the law doesn't bother me. It's the moralizing and the pungency they argue the position with."

Williams speaks about capital punishment only in an affectless, persuasive voice. He's a handsome, well-put-together man. If his clothes weren't quite so starched and fitted, and his grooming and manicure not so humorlessly exact, you'd be tempted to call him a dandy, but it would be a mistake. It's a presentation born of ambition rather than vanity, where everything from the silk socks to the off-tone collar speaks of success and a desire for more. The look is not atypical of a southern lawyer, and Williams is hardly the only DA down here who'll speak unambiguously of executions. Put that polish and fervor together, however, and Williams can seem quite cold-blooded at times.

Or perhaps it's just the decor. Ten minutes after entering the office, I'm still noticing lurid details. On the wall opposite his desk is a white plastic syringe inexpertly mounted on black cardboard, with stenciled letters spelling out Ulitmate Prick Award. The award is issued with each death penalty conviction in St. Francisville. "That's my first," he says. A larger syringe next to it, which has a big, painful-looking needle, is a gift from a friend. On the couch are two- by three-foot blowups used in prosecutions: One is a brightly colored "Abuse Chart," in which "Mitigating Factors" like "Intrauterine Drug Addiction," "Head Wounds," and "Sodomy Suffered as a Child" are color-coded and linked to the statistical probability of producing a murderer. Another is a jailhouse letter from Joel Durham, a young man Williams recently prosecuted for capital murder committed as a minor; Durham shot an assistant manager during a crack-crazed holdup of a McDonald's in the suburb of Metairie. The letter, presumably, was shown to the jury to prove that Durham was old enough and smart enough to be executed; the misspelled words, however, run about equal to those he got right, and the diction often makes the sense quite difficult to follow. Durham was given only a life sentence.

The decision, which was landmark because of Durham's age, was a painful loss for Williams. "Leo Kern, the guy Joel shot through the heart with his automatic weapon, had one ambition in life: to own his own McDonald's franchise. When this guy took vacation"—Williams's voice cracks with practiced passion—"he traveled to see the original franchise. I was trying to do Joel a favor," he says, the passion disappearing. "Pretty kid like him? That 'choir-boy smile' may have saved him with the jury, but it sure ain't gonna help him out up in Angola."

"You keep close tabs on your men," I say.

"You don't know the half."

Williams walks me over to a tiny replica of Angola's electric chair on his bookshelf. Pasted to its seat and arms are tiny cutouts of the faces of the men he's condemned. I notice he has one guy on there twice. "That's Saul Johnson." He smiles. "I got him for two separate capital murders, so they started calling him Electra-Saul—at least till the chair became cruel and unusual. They're all still on the row."

As we stand elbow to elbow at the bookshelf, Williams makes strong eye contact, the kind I can imagine him making with a jury foreman. I had mentioned Rideau's comment about the media's collusion in capital cases, and he'd shrugged it off. Now he wants to address it. "If I get you condemned, it's personal. It didn't happen because of some trick in the law, some exigency, some pressure that had been loaded up in the community or the media to make me prosecute you for a death sentence. It happened because I meant it to, and I want you to know it, in the worst way. When you finally go off, if you ever do, I'm going to be sitting right across that window while you're lying on the gurney. I want to be the last face you'll ever see."

"Have you considered moving to the appellate level yourself?"

"No, I like trying cases. I love it. The professional satisfaction is what you get from the victim's family, but emotionally, I have to say I love the battle itself. I'm a soldier."

"Jim's the fighter pilot," his colleague Ronnie Boddenheimer corrects Williams when he joins us. "I'm the soldier. I'm the doughboy."

Boddenheimer is a short, roughly groomed man who looks like a Dockers ad gone wrong—brushed khakis, cotton shirt, Timberland-style loafers, and a dull patternless tie that somehow manages to clash. You get the feeling he encourages the perception: a kind of Creole Lieutenant Columbo. Williams's voice is almost accentless, but Boddenheimer's speech is heavy bayou and infectious: when he and Williams start talking juries and Eighth Amendment issues, they sound like oystermen talking pirogue to pirogue. Boddenheimer isn't close to Williams's record, but the death sentences he has won, according to his abolitionist rivals across the river, are as ironclad as they get. "We're fighting a war here," he says. "There's moments of glory, but it's uphill all the way. For a country that's seventy, eighty percent pro-penalty, you wouldn't believe how hard it is to get your award."

"These people have jury-panel public relations firms working for them now," says Williams. "I'm talking public-defender clients with public relations firms?"

"You tell these people on voir dire you're going all the way," says Boddenheimer, shaking his head. "They smile. *Sure, sure, yeah, yeah.* They don't know how hard it is—at this point in history—to send a man to death. You don't want to bring race into it, but it's a fact of life. After *Gregg*, you could never get and would never want an all-white jury, but even if the murder was black on black, and most of them are, they still think it's white man's justice. But you gotta go racial neutral, or you're *going* to lose your award on appeal."

"Gender neutral, too," says Williams.

"Sometimes age neutral. You gotta *use* your dismissals. Never keep a teacher."

"No students."

"Never keep a social worker."

"Law students, especially."

"Unless he's studying neo-Nazi Techniques 101."

"We had one that made the jury was a Nazi. Remember?"

"Sure do. My choice for foreman."

"I have never cut a postal worker."

"Me neither."

"Psychiatrists, though . . ."

"Forget it. I believe they should be ousted from the jury pool altogether. Don't you?"

"You know I do."

I smile and scribble, though I realize, halfway through, that this interview is just character collecting, a little media collusion of my own. I take Rideau's point strongly, but it's instinctively clear to me that these men are not executioners, though they may well like to think of themselves as such. In Mississippi, I'd met two men who had actually put four people to death in the 1980s, and the difference between their somberness and the levity of Williams and Boddenheimer is all I need to see.

During a pause, I ask about Antonio James. Williams was still working for the New Orleans DA's office at the time of James's two murder convictions and remembers his jury coming out after the first. "The prosecutor let him have twelve women," he remembers. "Fatal mistake."

"So to speak," says Boddenheimer.

"They were smiling, too. When they're smiling you never get it. But he got it later. [James's death sentence came at a second trial.] Bye-bye, Antonio."

"Those days are gone," says Boddenheimer. "Man, I'll tell you. That chair was deterrence on four legs. Everyone knew what it did. Their eyeballs'd pop, they'd shit themselves. They'd roast and smoke and pee and burn some more. Don't let anyone tell you different, that chair was painful, that chair was hell, and people knew it. It was a deterrence, and now it's gone."

"Cheated Ronnie more than anyone," Williams says leadingly.

"What do you mean?" I bite.

Boddenheimer appraises me for a half second. "Sometimes at pretrial hearings, you get to stand next to the defendant," he explains. "Well, I'd shoulder up"—Boddenheimer leans in sideways, pursing his lips up toward me. His face colors, his upper lip puffs, and a slow whispering noise of electricity comes from deep in his throat. "*Bzzzzzzzz.*"

Jim laughs.

"Man, they'd bug. Eyes popping. Sitting in their Sunday suits."

"I think caning should be brought back," says Williams. "Publicly. And I think executions should be televised."

"Pay-per-view," Boddenheimer says. "If you're worried about ratings, get Howard Stern to pull the switch."

"I don't know how comfortable I'd feel making a circus out of it."

"Then I'll do it myself," says Boddenheimer.

"You'd have no problem pulling it?" I ask.

"I'd pull the switch and eat spaghetti," he says, then leans into me again. "*Bzzzzzzzz*," he says.

CHAPTER THREE

THE MISSISSIPPI GAS CHAMBER: A SPECIAL KIND OF PERSON

SEPTEMBER 1, 1983. For the first time in nineteen years, the four tiers of the maximum security unit at Mississippi's Parchman State Penitentiary were quiet. There were no mattresses burning in the cells, no toilets stopped with T-shirts to flood the hallways, no urine, feces, or fiery wads of paper flung at guards when they wheeled in meals, mail, and medications. The summer had been a particularly violent one for the maximum security unit (MSU), but as August drew to a close an entire week passed without a violent altercation: no one getting written up, carted off to the prison hospital, or dragged by four guards down to the "dark hole," a concrete bandbox at the front of each tier with a steel door on front and a hole in the floor for a toilet.

The usual cycle of threats, curses, and demands for attention no longer started the moment a guard or visitor appeared at the heavy double gates at the head of each of the four tiers. What was heard instead were the soap operas on the TVs of the thirty-six condemned men. Standing at the double gates, looking down the long, dimly lit tiers, all one saw were hands sticking out of every third or fourth cell, and an occasional gleam from inside these hands when the soaps went to commerical—flickers of light caught inside fragments of contraband mirror, or "peepers," the condemned's only way of seeing wider than what lay in front of the four-foot cell door. Inside some of the peepers one could see a pair of eyes, staring back at the double gates in an attempt to learn what was going on. Rehearsals for the asphyxiation of Jimmy Lee Gray had begun, and the sounds and odors coming from the long-unused death house—a red-brick annex at the end of C Tier—had made their civilizing impact, even on these men the state had deemed beyond redemption. After a nineteen-year moratorium, the death penalty was coming back to Mississippi.

If "the hangman's scaffold concentrates the mind," the gas chamber has a way of bewitching it. It's smaller than one would think, roughly four feet square and ten feet high. Almost beautiful, if one is mechanically inclined, it's also extremely alien looking, like an antique, six-sided diving bell someone painted gray, slathered with petroleum jelly, and jammed into a metal wall that divides two of the three rooms that form the death house.

Waist-high windows, tinted green and reinforced internally with thin wire, are embedded with large rivets in five of the chamber's six sides. At first sight, these windows make it seem harmless. Windows are hard to associate with death. Then the mind makes the obvious leap: this place is not only for killing but for offering death as a spectacle. Three windows look out from the rear half of the chamber onto the witnesses' room, where media people, state officials, lawyers, and families of the victims sit on long wooden benches that resemble church pews. A fourth window, on the right side of the chamber's front half, is for the two doctors who monitor the condemned's heartbeat on an EKG machine and a stethoscope. The fifth, to the left of the chamber's 300-pound door, is for the executioner.

The door itself makes the chamber seem very much a final destination. A meld of half-moon, concave pieces of ribbed steel, it sits on the left side of the chamber's mouth on a half-inch dead-bolt

hinge running through four immense casings, which allow no slack
or settling that would disturb the trueness of its fit. The oval doorway, which stands just under six feet, is rimmed with a rubber gasket caked with a heavy coating of Vaseline, to prevent any cracking
of the rubber that would allow gas to escape the seal formed when
the chamber is shut. A large submarine-style wheel on the door's
right is turned clockwise, several times, to secure the seal. The
wheel is so heavily oiled it makes no sound as it's turned. What one
hears instead is the oily slurp of the oval gasket gripping the door's
edge in the jamb, then a loud whirring as a small, squirrel-caged fan
kicks on in the ceiling, sucking up air to create the negative vacuum
in which the gas rises evenly. A heavy fizz of sulfuric acid, freshly
mixed with distilled water, comes seconds later, welling through
plumbing in a room below, followed by a heavy metallic noise as
the executioner drops the lever—it sounds like a silo door opening—
and a plop, as a pound of cyanide salts hits the sulfuric acid bath
below the chair.

There's a handsome brown wooden cross on the wall to the left
as you enter, a bare hundred-watt lightbulb in a metal cage above
it, and waist-high nooks in the three rear corners, which hold
beakers that look like the votive candles one sees in Eastern
Orthodox churches. The beakers are filled with an acetone-based
solution of phenolphthalein, a liquid that will turn scarlet when the
chamber's internal pH is greater than eight, indicating the poisonous gas has been neutralized with chlorides after asphyxiations,
and the chamber is safe for the living to enter. At the center, in front
of a thin ventilation pipe that runs from the plumbing room below
up to the death house chimney, is the squat, low-backed death chair,
which is surprisingly comfortable to sit in. Made of metal, its joints
have become indistinguishable under innumerable coats of black
paint. It has thick, high armrests, cupped at the back to secure the
elbows, and six black belts of good harness leather: at chest and
waist level, and one for each forearm and shin. The chair's seat and
side skirts are made of a wide steel mesh that allows the hydrogen-
cyanide gas (HCN) to rise unimpeded.

It comes up in a small cloud, then hovers like a phantom in the vaccum-sealed chamber until it's exhausted by ventilation and chlorine after the execution. Prussic acid, as HCN is commonly known, is colorless or pale blue in gaseous form, but in the harsh light of the hundred-watt bulb in Parchman's chamber, it comes up a dingy yellow vapor, which to some people smells like crushed or bitter almonds.

Cyanide is a small molecule that resembles oxygen—at least to mitochondria and other oxygen-receptive cellular components. It poisons by attacking the Krebs cycle of ATP synthesis, a phase of digestion in which six oxygen atoms are freed from the breakdown of each glucose molecule. In simple biological terms, cyanide tricks cells into absorbing it, rather than oxygen. Chemically expressed, it binds with and usurps the blood's available iron, inhibiting the flow of cytochrome oxidase, the terminal enzyme in the respiratory chain. Again, the oxygen needed to sustain life, though available in the blood, is prevented from being bound into the complex molecules formed in the acts of breathing and digesting, causing suffocation. In small doses (cyanide occurs naturally in peaches, apples, apricots, almonds, lima beans, chokecherries, sorghum, Johnson grass, etc.), it is rapidly absorbed in the intestinal tract, detoxicated, and excreted over a lengthy period. In the cloud directed up into the condemned's face by the overhead fan and negative vacuum, it forms what's known as the "chemical garrote": The lungs are free to breathe, oxygen is rife, but the man begins to strangle on his own breath.

The result is deep cellular confusion, a sensation not so much of choking as of oxygen starvation, triggering a prolonged but ineffectual panic response, which many observers of asphyxiations find the most difficult thing to watch. The lungs, responding autonomically to the adrenaline that comes with panic, begin to hyperventilate, causing more cyanide to be inhaled. The blood, deprived of the oxygen needed to produce hemoglobin, loses its clotting ability and flows poisonously into tissue, particularly of the lungs, brain, and the stomach's mucous membranes, which congest and redden, causing extreme burning and nausea. Rushes of blood to the brain, bringing increasing amounts of cyanide, cause massive disorientation. Cyanide, like many poisons, is a powerful hallucinogen. The lungs begin to hemorrhage and the throat and jaw constrict with the involuntary pressure that often precedes vomiting—a natural defense inhibited in the chamber by the increasing muscular dysfunction caused by deoxygenation. Instead, the neck area becomes both numb and tingly as the mouth fills with a foaming saliva, probably tasting heavily of acid and almonds.

Large patches of skin, deprived of oxygen, temporarily take on that purplish tinge known to coroners as lividity, the eyes bulge, and the pupils fix and dilate in an unnatural way. Condemned men whose eyes never close, or whose eyes close and reopen at a later point of the asphyxiation, which is common, seem to stare at a particular point or person. Large amounts of lactic acid flood through the body; accompanied by the massive adrenaline flow and oxygen starvation, the acid causes cycles of slackness and exquisitely painful muscle cramps, known as tetany. Under the loose work shirts, jeans, jumpsuits, or diapers the condemned wear, the muscles in the legs, arms, and shoulders are seen to undergo waves of unnatural flatness and freakish-looking tightening. It's as though the muscles themselves are suffocating and trying vainly to breathe, which in fact is what is happening.

This hyperventilation and panic is followed by a period in which breath becomes labored, then all but ceases. The condemned slump forward, in what's known as the "apoplectic form," and seem to dim out of consciousness as vital signs verge on flat and the heart rate falls to comatose levels. Though death is frequently declared at this point—generally one to three minutes into the execution—the heart, just as frequently, continues to beat, reawakening the body's defenses, at which point the EKG can record high and erratic rates. Condemned who have slumped thoroughly suddenly jerk their heads back, stiffen, open their eyes, and begin belching, farting, or gasping, seemingly returning to consciousness in a state of deep asphyxiation and florid hallucination as the battle continues—often for another four to eight minutes—before the lungs collapse and the oxygen-starved heart and brain finally fail.

It is a complicated, extremely indirect way of killing, unique among execution methods in that the condemned participate in their own deaths—and not just by breathing. HCN condenses at 78 degrees Fahrenheit, far warmer than a gas chamber's typical internal temperature, and the condemned's body heat is necessary to keep the cyanide gaseous for the eight to ten or more minutes before death is declared.

There are two schools of thought on how much of this violence is experienced consciously. Primitive animal studies, conducted not by scientists but by prison and state officials testing their chambers, have been inconclusive. Bedbugs, mice, rats, cats, and rabbits appear to die quickly and with no pain, but more complex animals last longer. While dogs, cows, and even horses were commonly electrocuted in nineteenth-century tests, piglets, for some reason, have been used in most of the eleven states that have killed by lethal gas, despite the fact that they seem to suffer for prolonged periods and squeal horrendously until they die. For humans, the period of panic and suffocation, designed to last no longer than five to ten seconds, seems, to the naked eye, to go on for minutes. Times of forty-five seconds to five minutes of consciousness were "recorded" by a San Quentin physician who monitored asphyxiations through the late 1950s and early 1960s. Though his measurements were probably just an empirical guess, he never doubted that the men suffered throughout. It will probably always be unknown if the condemned truly is sensate, both during the initial period and if and when he is reawakened, four to eight minutes into the asphyxiation. Firing squads, electrocutions, and hangings have all been survived and described, but no man has remained to describe the experience of being gassed in the chamber.

Thomas Berry Bruce, the executioner who mixed the chemicals or pulled the lever on the thirty-one men who died in the chamber from 1954 (when it was installed) to 1964, the year of Mississippi's last pre-*Furman* execution, never spoke of the men he watched die. He was a quiet man, who felt executions were something that had to be done but were not to be talked about. Even when his three children asked, he would answer in as few words as possible, and limit his comments to the chamber itself. To Bruce, it was a magnificent piece of machinery, and a far better killing apparatus than an electric chair, with which he had also had a great deal of experience.

In 1951, Bruce had become first a guard and then deputy to Mississippi's traveling executioner, C. W. Watson. The portable electric chair Mississippi put on the road in 1940, after the state switched from hanging, caused severe burning, disfiguration, and probably pain. Electrocutions were supposed to be private, but in reality there were sometimes dozens of people present in the county courthouses and jails where they were carried out, and gossip and lies about how the man had caught on fire, how his eyes had popped out, or how the blasts had knocked the lights out in the county courthouse were common. It had made perfect sense to Bruce when the state finally took executions behind the walls at Parchman in 1954 and switched to the chamber, which left only an unharmed-looking corpse and some detoxicated gas and gave reporters nothing more to write about than last meals and last words.

Bruce, the only American executioner whose career spanned the traveling days, the pre-*Furman* institutionalized executions, and the post-*Gregg* period, was born and raised in Belzoni, fifty miles down Highway 49W from Parchman. He had become a Belzoni patrolman in the late 1940s, between stints in World War II, in which he'd served overseas as a mess sergeant, and the Korean War, for which he'd been kept at bases in the South. He also moonlighted as a handyman and had a great love for "fooling with thangs," as he put it, everything from lawnmowers to cotton gins, no matter how used and rusty. He got the job as Watson's guard largely because of his availability—most men his age licensed to carry firearms in 1951 were in Korea—but he was handy with the chair and generator and quickly became Watson's deputy as well.

Mississippi's big oak electric chair and the diesel generator that traveled with it required constant maintenance, often needed most at crucial moments of the sentence—whether it was getting the thing to turn on at all or keeping the current at the required levels during the execution. Electrocutions were run by the county sheriff—he told Watson when to arrive, where to set up, and gave the order to begin—but all responsibility for the impromptu death chambers set up in the county jails or courthouses was Watson's. Bruce manned the 600 feet of cable that connected the chair to the generator, which stayed outside on the vehicle that the state provided—either a flatbed truck or horse cart, depending on local roads. He got the generator to kick over and reach a stable purr, helped shave the man's head and ankle, worked the conductive green gel into his forehead, answered his inevitable questions about how it would go, strapped him down, and set up the switchboard on a panel next to the chair. Watson threw the lever.

The generator presented great problems. Mississippi is a state of numerous mini climate zones, and it took a flexible hand with the choke to keep the generator running at speed—one month in the damp close of the eastern hills, the next in the Gulf Coast beach towns, then the delta. Humidity and temperature variances made for sudden slips and overdrives of current, both in the volts and amps, and the current needed to execute required three charges of differing voltages at a steady five amps. Bruce and Watson never learned how to ensure the exact charge. The condemned's eyes never popped (the tight-fitting headcap prevented that); the county courthouse lights went out rarely, if at all (generally, Watson and Bruce used their own generator); and the men never actually caught on fire, though they were severely burned in the chair. The initial blast, enough to cause a state resembling death, didn't cause the burns. The next two did. Any bodily fluid expelled by the heat of the first blast became a superconductor for the second and third. Saliva and mucous caused burning at the mouth, nose, eyes, and anus. Sweat, particularly in the areas under the electrodes at the shaved scalp and ankle, reached temperatures in excess of 140 degrees, causing a rapid cooking of the flesh, which at times would slough off the bone and begin to slide down the man's scalp and ankle; the sweat also generated steam that caused burns over the pores and orifices. Urine, if not as conductive as sweat, caused mutilation that, for obvious reasons, was difficult to behold.

The sight of murderers and rapists smoking, urinating, and writhing in the chair was what witnesses in courthouses and county jails of the early 1950s came to see. But Bruce was painfully aware of the families he and Watson had to send bodies back to. They hadn't committed the crime their father, son, or brother was punished for, and there was no need to inflict the sight of a charred body on them. Burns, sores, and odors also reflected badly on his professionalism, which meant a great deal to him—he was a perfectionist.

Those problems came to an end in the fall of 1954, when the state got rid of the electric chair, switched to gas, and conducted all executions behind the walls of the state penitentiary. Bruce and Watson went up to Parchman to learn how to run the chamber being installed in the death house of the maximum security unit, the new building put up that summer to house Mississippi's first death row, as well as the prison's worst inmates. An engineer for Eaton Industries, a Salt Lake City company that built ten of America's eleven chambers, ran them through the drill: Bruce, the deputy, would mix up the sulfuric acid solution in a separate chemical room; Watson, the executioner, would place the cyanide under the chair, throw the switch, then detoxify the chamber afterward and clean the body. The Eaton man had been instructing executioners since 1938, when California switched from hanging to the gas chamber.

At Parchman, Bruce realized, he was no longer just a deputy: his role in mixing chemicals provided half the coup de grâce. His new title, inexplicably, was sergeant of security. As a guard in the traveling days, he had been responsible for Watson's security, escorting him and the chair. At Parchman, he wasn't allowed to carry a gun, and he no longer escorted the executioner. In fact, Bruce was now to be chauffeured to and from executions himself, by his daytime boss, the Belzoni sheriff.

Watson remained the executioner, and received twice Bruce's pay for it. He even had a certificate, given to him in Jackson, that named him "Executioner for the State of Mississippi, Serving at the Pleasure of the Governor." The wording dated back to a clause that had been tacked on to the 1938 statute enacting the state's move from the gallows to the electric chair, and reflected an even older tension. The relationship between Mississippi's legislature and governor had been strained since antebellum years, when the executive branch had been loaded with Union carpetbaggers put in Jackson to effect Reconstruction. For sixty years, Mississippi's governor had putative executive powers, but the legislators owned the state, and it behooved them to keep things regional and backward.

That included executions, which until the late 1930s were public hangings, gala affairs conducted by the county. The traveling electric chair had been a shrewd move, made by a new breed of centralists who had come to Jackson in the 1930s, to exert power over the counties. They tried initially to centralize the death penalty in Sunflower County, passing legislation that mandated all executions to be conducted at Parchman. Both the Parchman warden and Sunflower County legislators objected strenuously, refusing to be known as the "Death County," and the compromise of a traveling electric chair was reached in 1939. After some difficulties in finding a manufacturer, the chair was finally delivered by a Memphis engineering firm to Jackson in 1940, and was put on the road on October 11, 1940. The condemned was a "little, sawed-off, tar-colored wife-killer named Willie Mae Bragg," as a contemporary account defined him. Initially run by an ex-con/former traveling somnambulist named Jimmy (or Jimmie) Thompson, who had performed on the carney circuit under such names as Dr. Alzedi Yogi, Dr. Zogg, and Dr. Zingaree, the traveling chair remained the method of choice for fourteen years, though the condemned still had the option of a simple, county-held hanging.

Putting all executions behind Parchman's walls in 1954 was the second leg of that centralization, and the gas chamber was the final touch. The counties objected to having their right to execute taken away, but the fact remained: you can't put a gas chamber on the road, and the state executioner, serving at the governor's pleasure, was licensed only to execute by asphyxiation.

Mississippi's first asphyxiation was of Gerald Gallego, a Biloxi man who had slit the throat of one of his jailers during an escape attempt from the Pascagoula County jail. Bruce was probably curious to see how the new chamber would work. It was March 1955, and the Gulf Coast in early spring had always tested their finicky generator. With traveling, the job would also have entailed a long two days on the road.

Executing a man off death row was entirely different from doing so in the various county seats. The sheriff of the county in which the condemned had been convicted still traveled to Parchman to "conduct" the execution—in a ceremonious nod to the old ways—but he was now a complete figurehead, and generally as awestruck by the chamber as everyone else when they first saw it. At Parchman, Bruce and Watson had control. Witnesses, limited to twelve, were a complete irrelevancy to Bruce, who saw them only through the green-tinted glass of the gas chamber and no longer had to deal with their senseless gossip and questions. In Parchman, there were eight "witnesses" to Gallego's execution, whose presence in the MSU made great sense to Bruce—Mississippi's first death row population. These condemned men didn't see the asphyxiation, but the new death house was just the other side of the door from C Tier, meaning the closest condemned was two cells away. Gerald Gallego walked to the chamber with a priest saying the Lord's Prayer, as the men always had with the electric chair, but here there was a chorus behind him in C Tier: "Up Above There's Heaven Bright," sung by eight men whose turns were coming. A year earlier, those men would have been sitting in eight different county jails.

Watson dropped the lever on Gallego and nothing happened. The lever lowering the cyanide crystals into the sulfuric acid bath hadn't dropped, and Gallego had to sit it out in the chamber while Bruce went in and fixed it. Watson dropped the lever again, and this time the crystals submerged easily beneath the chair with a quiet wisp. A bit too quiet: only a handful had made it into the acid bath, and the gas that came up was sufficient only to sicken Gallego. He was conscious as Watson decided to evacuate the chamber and start over; it took another twenty minutes to detoxicate the gas, get the chamber unbolted, put another pound of cyanide below the chair, close the chamber up, and start again. A full half hour passed before the attending doctor could declare death.

The witnesses were ushered out of the death house, and after the phenolphthalein beakers turned scarlet a second time, Watson put on vinyl gloves, a leather apron, and an oxygen mask and went in to dust Gallego down. The only mark was on his jeans, where he'd wet himself. Watson ran his gloved hands meticulously through Gallego's hair and the folds in his clothes to shake off the residual cyanide particulate, then he and Bruce took him out of the chamber, stripped him, hosed him down, and dressed him for the funeral home. Compared to the disfigured bodies that emerged from the electric chair, it was like washing a newborn. There was no shame in handing a corpse like this back to the family.

Running a gas chamber, however, would prove far more problematic than Bruce and Watson had been taught to expect. Preparing the sulfuric acid bath for the cyanide crystals, for example, was an extremely tricky business. Even with distilled water as a solvent, the admixture had a treacherous way of backlashing up any plumbing system it was poured down, and it bubbled unpredictably, releasing poisonous vapors that, while not immediately lethal, had a cumulative effect. The cyanide liquid in the drip pan below the chair stayed deadly for two days, and the long sewer line that took the effluent down to a lagoon had to be checked for leakage into the aquifer. Though the gas that left the death house smokestack was harmless (witnesses would see birds hovering blithely above the chimney when it came out), the ammonia chlorinated only the airborne poisons above 78 degrees. Otherwise, the cool air inside caused the lethal particulate that covered everything in the chamber: if an asphyxiated man's body were handed to a funeral home without being dusted, stripped, and hosed, the residue leaching from his pores would kill the undertaker on contact.

Watson died four years after he and Bruce started working at Parchman. The contemporary news reports mentioned a chronic respiratory illness, and suggested he had been felled by sulfuric acid fumes from the chemical mixing room. This is unlikely, since Bruce did the mixing, which happened behind closed doors in the chemical room, to the left of the chamber, and both wore oxygen masks or gas masks the entire time the chamber was in use, even during dry runs, when they used placebo chemicals. After Watson's final execution, of a Hattiesburg man who had beaten a woman to death for a purse that contained a nickel, Bruce took over at Parchman. He hired the brother of his best friend, Hoop Farmer, as his sergeant of security, and several years later received his own certificate. Bruce loved the phrase "serving at the pleasure of the governor."

He stayed busy. After leaving the police force, he held numerous jobs in Belzoni, which is a bit more thriving than the typical delta town: Catfish Capital of the World, they call it, because of its large fillet refinery. It has a sizable Jockey plant, where underwear is made from the local cotton, and Bruce eventually became Jockey's chief of maintenance. He also worked as a millwright, became the custodian of the public school, and sold vegetables at a roadside stand at the crossroads of highways 49W and 12 on weekends and holidays.

Roughly four times a year, he'd get a call from the Parchman superintendent, and let his supervisor at Jockey know he might be leaving early one day a few weeks hence. Then a morning would come with a story in the paper about another execution at Parchman, and Bruce would show up a little late, tired and quieter than usual. People at the plant knew better than to ask, and not because of unease, or distaste for the work he did. Bruce received the occasional piece of anonymous hate mail, accusing him of working for "blood money," but when he brought his wife and kids to the Belzoni Baptist Church on Sundays, he knew there wasn't a soul in there who harbored contempt or thought him notorious.

Bruce was a man of parts. As the years passed, he became increasingly quiet, particularly about his work at Parchman, but he was also known as one of the local characters, and people would stop by his roadside vegetable stand just to savor the pleasure of his company. His first wife died of cancer, and when Bruce remarried during the moratorium on capital punishment, he didn't tell his second wife for many years that he was the executioner. When he finally did, she was shocked that her husband had carried out the state-sponsored murder of his fellow man. That was exactly how Bruce saw his work: on the death certificates he signed as a member of the six-man coroner's jury, the box checked for Cause of Death was Homicide, which was literally what had occurred in the chamber. To Bruce, it was like *pesticide* or *herbicide*. It was just a word.

Like many executioners, Bruce's lethal squeamishness was instead for the sight of an animal's pain—even of a fresh-caught carpy or catfish. He grew up hating hunting, and as an adult gave it up altogether. He still went fishing at Townsend Lake, but he rarely came home with anything. Instead, he relied on friends to bring him fish and game, and many did. Bruce was a good cook, who knew when to take food off the fire, whether it was a long-simmering venison stew or a quick plate of catfish, rolled in corn-meal, spiced just right, and shallow-fried in fat that was always the right temperature. He was a good gardener, with a green thumb for all manner of fruits, vegetables, and nut trees. The vegetables he sold at his stand were the best for miles around.

Bruce had a local reputation for violence as well, but it didn't come from his work at Parchman. It came from Saturday night at the bar, where he had a way of getting mean and indignant, and where someone would inevitably get up the pluck to test his famous right hand. Bruce got particularly quiet when he drank, chain-smoked Luckies, and kept to himself at the end of the bar. During the week, he'd have a beer or two after work, maybe more, but on Saturday nights he drank whiskey. In fights, he wasn't quick or strong so much as decisive. Before you knew you were in a fight, you were on the floor, and Bruce was back at the bar where he'd left his drink and his cigarette in the ashtray. On occasion, he drank too much and started up with two or three men, and he'd take a beating. His friend Hoop Farmer remembers Bruce taking some bad ones. "Berry fought more and more the older he got," Farmer says. "Supposed to go the other way, I thought."

No one, including Farmer, ever learned why Bruce had taken the executioner's job, or why he'd kept it all those years. It wasn't conviction or pride in the title. He had an opinion about the penalty (he was for it), but that was as deep as it went. If he had any feelings about retribution or views on the politics of capital punishment, he never shared them. It may have been simply the money, but the chamber was no sinecure. Over three decades at Parchman, his fee per execution rose from an initial $75 to $250, but the work included dry runs and maintenance of the three-room suite: upkeep of the plumbing, meters, and chair, lubing the apertures, surfaces, and gaskets, then testing everything for leaks with lit candles at the end of the day.

That work may have been the appeal to Bruce. Even during the state's nineteen-year moratorium on the death penalty, he appeared once or twice a year, unbidden and unpaid, at the double gates of the maximum security unit. The guard walking the parapet at the unit's pink tower would lower a red plastic bucket and, like all visitors, Bruce would surrender his keys and wallet, yell up answers to rote questions about firearms he was carrying, and wait for the gate to slowly open. Then he walked to the rear of C Tier, opened the locked door to the death house, spent the day "fooling with" the chamber, lit his candles and tested it, and swept the place out.

After *Gregg* came down, Bruce took on a death house apprentice named Donald Hocutt, a twenty-two-year-old MSU guard with an unusual willingness to take on the prison's more unpleasant jobs. Then fifty-seven, Bruce was a taciturn, matter-of-fact teacher, who had an unfiltered cigarette going between his lips seemingly at all times and an amazing capacity to smoke an entire butt without tipping or losing the ash. He had no stories about the old days, only curt instructions about the tasks at hand.

Hocutt had been hooked since he'd laid eyes on the chamber a year before, glowing with its lights on in the dark, unused death house like a spaceship from a "Buck Rogers" TV serial. There were eight trustees at the MSU, working in the kitchen or the laundry behind C Tier. The oldest, a lifer, had showed Hocutt around the chamber a week after he'd started at Parchman. The trustee had worked on it for years, helping Bruce with the upkeep both before and during the moratorium, and said he had even made some money on it. In 1968, the deejay of the local rock 'n' roll station put on the Rolling Stones' single, "Get Off My Cloud," and offered $25 to the listener "calling in from the strangest place." The trustee, presumably calling from the "hotline" phone inside the chamber, won easily.

Hocutt knew from movies that cyanide pellets went under the chair, but he couldn't see right away how they got in the chamber without going off and killing the executioner, or how the sulfuric acid made its way to exactly the right spot to dissolve the crystals. Hocutt came from a family of mechanics—more accurately, failed farmers who by default had become fix-it men for the county. After the bulldozers, backhoes, and tractors he'd run since the age of eleven on his family's farm in Garland County, Arkansas, he had a sixth sense for machinery he'd never seen before. Try as he might, he couldn't figure out how a poison so deadly could be created without risking the executioner or the witnesses, how the chamber could be evacuated safely, or how they knew when that point had come. Trying his hand at the lever to the left of the massive door, however, he was fascinated by the thought of what it would feel like to do so with a man in there.

He walked into the chemical room to the left of the chamber. It was a dark, unpleasant room with one small window, but Hocutt felt at ease inside there immediately. Garland County, where he'd come of age, was a dry county, and in his early teens he had delivered whiskey and strong beer to the Texarkana honky-tonks, running it in gallon plastic milk drums from a still his uncles ran in the backwoods of a local politician's huge estate. The dark enclosure of the chemical room and its oversize stainless steel sink and plastic jars and bottles reminded him of the backwoods still, and the quiet of the room also appealed to him. After the violence and screaming on the tiers, less than thirty feet away, it was peaceful in there.

It had been years since sulfuric acid had been mixed in the chemical room sink, and it took a few visits to figure out the process. The chemicals were mixed in the sink itself, which was then opened with a valve that allowed the acid bath to run through the tubing below, in the plumbing room, then across into the well below the chamber seat. There was a trapdoor in the floor of the chemical room down to the plumbing room, and Hocutt went down for a long look at its immaculately maintained copper pipes and freshly painted ventilation shafts. It was fifteen degrees cooler, and the air was crisp, dry, and healthy-seeming. Looking from below at the mechanism by which the lethal gas was unleashed, and seeing how much care had been taken—how every gasket, valve, and joint was perfectly lubed and sealed—it was clear to Hocutt that this was the domain of a special kind of person.

CHAPTER FOUR

THE EXECUTION OF JIMMY LEE GRAY

JIMMY LEE GRAY came to the maximum security unit in early 1977, dripping chains. Like all prisoners sent directly to the MSU from jails outside the penitentiary, he was transferred to Parchman inside the "bullpen," a gated area of the MSU's long main hall where convicts waited, alone and shackled, while their paperwork was seen to.

Gregg had come down only nine months earlier, and Gray was the first condemned man Donald Hocutt had seen getting checked in. He looked anomalous in the bullpen. The guys Hocutt had seen in there usually looked like the worst kind of trouble, but Gray was just a nondescript, unpleasant-looking man. Of medium height, he had brown, narrow-set eyes that by turns made him seem cocky, absent, or paranoid, and small, bone-white hands that looked womanish. His face had the pasty, unhealthy complexion men can get after long imprisonment, and his dark, longish hair, worn in a ponytail, was greasy. After twenty long months at Parchman, Hocutt felt he had gotten to know the various criminal types, but Jimmy Lee

Gray had a special perversity, an unsavoriness that made Hocutt want to get as far away from him as possible. He was a pariah.

At twenty-eight, Gray, who had grown up along the California-Arizona border, had been in jail all but fourteen months of his adult life. He had a strange criminal history, marked by sudden outbursts of violence, followed by vague attempts to cover his tracks and quick confessions. In 1968, at age nineteen, he had committed his first murder—strangling and slitting the throat of Elda Louise Prince, his sixteen-year-old fiancée, whose family had taken Gray in to live with them. Gray hid Prince's body in a culvert, then led police to it the moment he was questioned about her disappearance, saying she'd made taunting remarks about his sexual prowess. Sentenced to twenty years by the State of Arizona, he got out after seven years of good time, during which he'd learned to program computers.

He was hired by a Chicago firm, which had an opening at their office in a Pascagoula shipyard. In June 1976, fourteen months after Gray moved to Mississippi, a three-year-old girl named Daressa Jean Scales went missing from the apartment complex where he lived with his girlfriend. When a computer search for criminal history among the complex's residents turned up Gray's name, he was brought in for questioning and quickly admitted he had invited the little girl up to his apartment to see his kitten. An hour later, he led a Pascagoula detective to a woodland creek off a logging road thirty miles away, where the girl's body floated near a small bridge. Gray claimed she had fallen off while playing, but when an autopsy revealed she had been raped and sodomized and that she had mud in her lungs, he confessed that he'd held her face down in a bog created by a drain-off from the creek, then carried her body up to the bridge and thrown it off. He was convicted in a short trial in December and sentenced to die in Parchman's chamber a month later.

Thomas Berry Bruce was at work in the chamber room less than a week after Gray's arrival, but as the execution date neared, it was clear that it was going to take longer than a month to get a post-*Gregg* sentence carried out. The date passed and Gray's name drifted off the front pages of the Jackson and Pascagoula newspapers, until a story appeared in the Jackson *Clarion-Ledger* about letters that Gray's mother, Verna, a former mental patient, had written to Governor William Winter and the Mississippi Supreme Court. In both, she pleaded that her son *not* be spared, for she felt he "deserved to die." An AP reporter tracked her to San Diego, where she had joined a victims' rights group and devoted her time to lobbying for capital punishment laws in the State of California. "I love him," she said of her son. "I guess I love him; I'm not really sure. . . . I hope he finds peace of mind in the next world. He hasn't found it here."

Articles about Gray's childhood began appearing, with details filled in by depositions of family members and former teachers that were taken by volunteer appellate lawyers trying to keep him out of Parchman's chamber. Gray's father had left his mother to marry her best friend when Jimmy Lee was less than five years old, and his mother had taken to drink and become suicidal. She sent Jimmy Lee to live with his father, in hopes it would destroy the new marriage, but he'd soon come back to her, beginning a hellish ten years of abuse and neglect. He spent the decade shuffling between his mother and an elderly great-aunt, who testified that Verna Gray had "the face of the Devil" when she flew into rages, and that Jimmy Lee was "a sweet boy with a split-personality . . . quiet, aloof, gentle, until something set him off, after which he'd always apologize real quickly." At school, he was recognized equally for his high IQ and explosive reactions to assignments he didn't like, fits of temper that were followed by apologies and long spates of good behavior.

In April 1978, the Mississippi Supreme Court overturned his conviction for the murder of Daressa Jean Scales on technical grounds and rescheduled a second trial. Gray pleaded insanity this time, but court-appointed psychiatrists found him competent to stand trial. One noted that Gray was emotionally disturbed—"He gets very angry, very easy," he wrote—but nonetheless remained aware of what he was doing at the time. Gray was again sentenced to die, with a date set for June, two months later. "Jimmy Lee Gray," wrote the judge who signed the death warrant, "is just plain mean."

Gray made no friends on death row, and was never heard in conversation between cells. Being a child murderer didn't win him any friends (many men in the MSU had been beaten or molested as children), but Gray had no disciplinary write-ups and only one fight, a momentary flare-up while in transit—a remarkable record given the MSU's violence. The men seemed simply not to want anything to do with him. The only mail he received was from his lawyers in Mississippi and Alabama, advising him of various appeals. There was a flurry of correspondence in the second week of May 1978, when the United States Supreme Court considered his sentence and stayed his June execution date. Then the case went back to the state appellate level for nine months. In the early spring of 1979, the Mississippi Supreme Court, weighing an appeal based on Gray's abusive childhood, finally ruled that "Gray committed a brutal killing for a sordid purpose of a helpless child," and sentenced him to die on Halloween night, 1979.

Three other men received earlier execution dates. The first was Charles Sylvester Bell, a twenty-two-year-old Hattiesburg man who had murdered a service station attendant a few weeks after *Gregg* had come down. His date, April 12, 1979, came up more quickly than anyone in Mississippi had expected. John Spenkelink, in Florida, was due to go six weeks after, and throughout the Death Belt it had been assumed Spenkelink would be the first. The sudden urgency of Bell's date made Parchman's wardens worry about the state of their chamber. Bruce had assured them it was in fine condition, but they wanted a second opinion. Calls to Arizona, North Carolina, Nevada, and other states with chambers were no help, and they finally called Eaton Industries in Salt Lake City. Two Fridays before Bell's date, Hocutt and his friend and fellow guard Ronnie Fulcher drove to the airport at Memphis to pick up the engineer who had instructed Bruce and Watson in 1954. Now in his seventies, the man was given Parchman's guest house, next to the warden's house on the prison's main road. He stayed the weekend, spending every daylit moment in the death house, deciding which gaskets, valves, belts, meters, windows, laths, tubes, filters, and pipes had to be replaced or oiled, relined, or recalibrated. Hocutt rarely left his side.

Bell received a stay from the Fifth Circuit Court of Appeals in New Orleans, and later made it off death row into the general prison population. A month after his stay, the dates that had come up for two other prisoners were also stayed by the Fifth Circuit. In September, the Supreme Court stayed Gray's Halloween date and his case went back to the state.

Hocutt, who became lieutenant in charge of the maximum security unit during Gray's third round of appeals, was convinced that no one would ever go to his death at Parchman. He went on maintaining the chamber, however, and became the resident expert on asphyxiations. Jimmy Lee Gray became a born-again Christian and began writing poems. A reporter doing a story on death row even managed to get a few words out of him. "There are no flowers here on death row," he said, adding that he would write her a poem about flowers. In 1981, a first-prize certificate came from a magazine for a poem about flowers that Gray had written to an eight-year-old girl who'd sent him a Christmas card.

He also began to write letters, though not to his family, the families of his victims, or even the appellate lawyers who had been keeping him alive since 1977. They went instead to a lawyer handling the divorce of Daressa Jean Scales's parents, whose marriage had failed in the aftermath of the murder. "There are a lot of people like me in this world, but nobody really cares . . . until it is too late," he wrote in one. "The only solution they can come up with is give him the death penalty. At least that's the solution that satisfies angry people." After ten years of imprisonment, Gray was still living in the psychological yesterday of his own parents' divorce.

During his first five years at Parchman, Gray had refused to see the church ladies and pastors who came every Sunday to visit the condemned. That began to change as his appeals inexorably wound down. By 1981, his case had been pled before over twenty judges; by mid-August 1983, it would appear before fifty-nine, including three Supreme Court hearings. The first visitors he agreed to see were the women from the local Baptist church. Then, elderly representatives from the Pilgrim Rest congregation outside Indianola came to offer him a gravesite, should he ever get executed. They soon began to come from greater distances: a Natchez man, who told Gray he had no feelings about the penalty but felt Gray needed a friend who wouldn't judge him; a Holy Roller from Greenville; the Reverend L. C. Dorsey, a famous abolitionist with the NAACP in Jackson; and a pro-penalty minister named Don Dickerman, associate pastor of the Shady Oaks Baptist Church in Hurst, Texas, who was attracted less by Gray's devotion to Christ than by the nature of his penitence. "Jimmy has an unusual philosophy," he told a reporter. "He feels he deserves to die according to the law of the land"—a sentiment echoing the views of a prominent televangelist, who had claimed that Jesus Himself had shown his support for the death penalty by submitting to crucifixion, when as the son of God he could have gotten out of it.

In July 1983, Gray had a close brush with the chamber when the Fifth Circuit turned down a challenge to the constitutionality of Parchman's chamber, on grounds of cruel and unusual punishment. It wasn't the first time the issue had been raised on appeal, but it had been thoroughly researched by Dennis Balske, Gray's new lawyer from the Southern Poverty Law Center in Birmingham, Alabama. The appeal attracted a great deal of attention and put pressure on the three justices overseeing executions for the Fifth Circuit; they devoted long study to the affidavits and medical reports Balske had submitted.

Their final ruling, which came amid eleventh-hour telexes and overnight mail, was extremely direct. "We are not persuaded that under the present jurisprudential standards the showing made by Gray justifies our holding that as a matter of law, the pain and terror resulting from death by cyanide gas is so different in degree or nature from that resulting from other traditional modes of execution as to implicate the Eighth Amendment right [forbidding cruel and unusual punishment]." That fifty-eight-word judgment was followed by a far more straightforward one: "Traditional deaths by execution have always involved the possibility of pain and terror for the convicted person"—probably the frankest words ever issued by a modern court on the punishment.

It was late June then, and Gray's newest execution date was less than two weeks away. His lawyers filed petitions with any court that had a say in the matter, actions that seemed increasingly like Gray's final legal death throes. Superintendent Eddie Lucas, Parchman's warden, asked Thomas Berry Bruce to get the chamber ready again, then put the penitentiary on institutional emergency, barring outside visitors and locking prisoners down in their cells. Hocutt had Jimmy Lee Gray moved from his cell on death row to the last cell of C Tier, adjacent to the death house, and put him under a twenty-four-hour death watch.

With less than a week to go, a stay came down from the Fifth Circuit, which was besieged with appeals and last-minute decisions. A watershed had broken, and a huge number of cases had reached their crucial points concurrently. With motions flying simultaneously in every southern state to every court imaginable, the Supreme Court decided to hear the appeal of what was to become a landmark Texas death sentence—of a murderer named Thomas Barefoot, who argued that the Supreme Court alone should have final jurisdiction on the penalty. The Fifth Circuit didn't want Gray to die until the Court's opinion in *Barefoot* had given them clear precedent.

Hocutt had Gray taken back to his cell. "If this is God's will, then okay," Gray told Parchman's chaplain, Reverend Ronald Padgett. "I wish they'd hurry up and get it over with."

The Court came down against Thomas Barefoot two days later, ruling that the burden of execution stays lay with regional federal appeals courts. The Fifth Circuit, the regional court for the heart of the Death Belt—Mississippi, Louisiana, and Texas—promptly passed on Gray's rescheduled date of midnight, September 2, 1983. He had no appeals left, and his only hope was with the governor's office in Jackson or a change of heart from the Supreme Court.

•   •   •

Hocutt had Gray brought back to C Tier on August 25, though he had no belief Gray was going to die in the gas chamber on September 2. He had been receiving photocopies of death warrants, orders, and execution-date memos since he'd assumed the MSU lieutenancy in 1979—from clerks at the Jackson district attorney's office, the state supreme court, the Fifth Circuit—as well as telexes from a special death penalty clerk at the Supreme Court. When Gray's July date had begun to seem possible, Hocutt had ordered two phones installed in his office, as backups to the two already at the death house. There was a story making the rounds again at Parchman, which was a terrible gossip mill, that a sentence in the early 1960s had been carried out by accident despite a last-minute stay from Justice Earl Warren. Hocutt had been hearing that story since 1976, and knew it wasn't true, but he wanted backups for what he regarded as the inevitable last-minute reprieve.

The men on death row, oblivious to these details, seemed to know on August 25 that this time around was different. As Gray made his third "last walk" to C Tier in as many months, shouts of "Be cool" and "Hang in there, Jimmy" came from all four tiers of the unit. Despite his unpopularity, the shouts continued for hours after the cell's solid-steel door was shut behind Gray. Then a week went by in silence. The only noises heard throughout the day were of the five-man execution team: lieutenants Donald Hocutt, Ronnie Fulcher, and Bobby Gregg and the penitentiary's ranking officers, majors Fred Childs and Roger van Landingham. They began preparations warily, feeling silly in their black latex gloves, gas masks, and yellow rubber aprons, but the initial strangeness and anxiety wore off quickly. Death was not new to any of them, particularly as a team. In a hierarchy predicated largely on cold and efficient brutality, they were Parchman's elite, and had worked together for years on riot calls, dealt with hostage situations, chased escapees across the prison farm, and hunted down those who had made it outside the fences.

After the first practice asphyxiation, they began to make jokes, usually at the expense of the officer sitting in that day as Jimmy Lee

Gray. (For each rehearsal, one of the officers was led to the chamber, strapped in, and read a death warrant.) The wife of the unit's chief of security, Barry Parker, sent batches of fudge brownies in the afternoons, and the men nibbled from the trays and drank Cokes while keeping charts of the procedure's thirty-seven steps. Starting with a click of the stopwatch, they shut the heavy overhead windows between the death house and C Tier and behind the witness room, then led the stand-in, who had a heart monitor taped to his chest and got no brownies, to the black metal chair. Strapping him in, they poured a pound of salt into the cone-shaped dispenser below the chair as two others brewed placebo acid baths in the chemical room. Then they sealed the chamber, checked the negative vacuum on a gauge to the right of the door, threw the lever, and lit candles to test every juncture from which air could escape. If Hocutt reminded them they were at a potentially dangerous moment, someone invariably deadpanned, "Someone could get killed in there." And when the small ceiling fan kicked on with its surprisingly loud roar, it got a rise from the man in the chair—you couldn't help but shrink in horror from it—and everyone outside the chamber would howl, even the one who'd been screaming in there the day before. Whenever the laughter grew too loud, Hocutt would nod to the wall of the chamber room, behind which sat Jimmy Lee Gray.

On August 28, four days before the execution date, the five men tested the chamber with real cyanide and sulfuric acid on two rabbits Hocutt had had the K-Nine squad round up. Hocutt, standing in for the strap-down team, entered the chamber and put the rabbits on the metal chair, feeling ridiculous. Resisting temptation to try for a laugh by reading them a death warrant, he stepped out backward, then the door was shut and bolted. Half the team performed the execution; the other half sat in the witness room, to ensure that no gas leaked from the rear of the chamber. The cyanide hit the sulfuric acid with a loud plop and the rabbits scratched their noses as the lethal gas came from below the chair. Ten seconds later, they fell over on the chair, their tiny legs pointing straight up.

Two days later, Hocutt had the K-Nine squad round up two more rabbits. They were put in, he threw the lever, and again their little legs went up. This time, however, Hocutt and the two other officers running the execution, Gregg and Childs, began to grasp at their throats seconds after the gas came up. Stumbling to the ground, they went into convulsion, their arms writhing, their own legs sticking up. The three officers in the witness room tore out screaming through the chamber room and the back door of the death house, their faces covered by hands, arms, and shirts. Twenty paces later, they stopped and looked back at the death house door, where Hocutt and the others were falling over each other and laughing themselves sick.

On August 31, a Wednesday, and the final day of practice, the team was joined by Bill Raimand, a commissary officer who'd volunteered to fix Jimmy Lee Gray's last meal in the MSU kitchen, next to the laundry room behind the death house. Everyone was glad to see Raimand, who had worked with them for years on the emergency rescue team, chasing escapees.

Condemned men tend to request meals that evoke childhood. Gray, who grew up in dusty, agricultural areas of California and Arizona, asked indifferently for "a plate of Mexican food" and salad, then also asked for a glass of milk and a bowl of strawberries. Raimand put every imaginable form of tortilla on the plate, plus rice and salsa, and peppers and olives in the salad. The smell of edible food in the MSU brought the trustees out of the laundry like a magnet. The old man who'd shown Hocutt the chamber in 1976 had told them that the condemned never finish their last meals, and that there should be plenty for everyone after Jimmy Lee went. Gray, who had given his last meal request to Deputy Warden Joe Cooke of the death watch team, had asked that the menu not be revealed to the press, but the trustees watched every pot on the stove boil. Gray had had no other requests or last words, except that his burial site also be kept secret.

Because the execution was scheduled for midnight, Friday morning, and the following Sunday was Fall Rodeo, which was

always closed to the public, Superintendent Lucas ordered the prison
closed on late Wednesday to all nonstate outsiders, save the press
chosen to attend the execution. He also put the population on lockdown again, which seemed like overkill to the execution team, just
as it had when Lucas had made the move in July. It was like something he'd seen in a movie, and Hocutt had enough of that feeling as
it was. Everything he'd been doing for the past week seemed unreal.

He awoke early on Thursday, feeling like he was falling off the edge
of a cliff, astonished by how fast time was passing. Governor
William Winter had announced after meetings with abolitionist
ministers the day before that he would not grant Gray clemency,
and any thoughts of a last-minute change of heart were canceled by
what Hocutt heard on the radio: Winter would be spending the
evening at a concert by Indianola's own B. B. King at Ole Miss.
Oxford is a two-hour drive up I-55 from Jackson, which meant that
the governor would be on that road at midnight, should conscience
strike him. Hocutt and the rest of the team had been invited to
share any of their own apprehensions with Reverend Padgett, or
Charlie Jones, the prison's Southern Baptist minister. After showering,
Hocutt went to speak briefly with Jones, whom he liked and trusted.
He was offered little but homilies—this was God's will, and Hocutt
was only an extension of that will and of the will of the people—but
it was enough.

Thomas Berry Bruce arrived at the MSU gate at 2 P.M., wearing
a plaid shirt, beige trousers, and black shoes with white tennis
socks. The wife of his longtime deputy, Mr. Farmer, had taken ill,
and though Farmer had never missed an execution before, he'd
begged off this one. Bruce had instead brought another Belzoni man,
named Mr. Jones, who seemed nervous. Mississippi's laws explicitly
and implicitly mandated that nonprison outsiders conduct executions, but the absence of Bruce's regular deputy meant, de facto, that
Hocutt would be mixing the sulfuric acid. He was the only man in
the state with any experience of the chamber besides Major van

Landingham, who was in charge of the strap-down team. With a total lack of ceremony, Bruce deputized Hocutt and Mr. Jones, then walked them through the chamber and chemical room, telling Jones just to follow Hocutt.

By tradition, Parchman's superintendent, Eddie Lucas, had absented himself from as much of the execution protocol as possible, and he continued to do so on the final day. Lucas would normally have entertained distinguished visitors to Parchman such as Bruce, but as always on execution dates, Bruce went for supper with the ranking member, Major van Landingham, who after supper issued Bruce the cyanide crystals from a vault in the armory. Then Bruce spent the rest of the day by himself in the guest house until the Humphries County sheriff, who had driven him up 49W that afternoon, brought him to the death house early in the evening. By some accounts, Bruce did a bit of drinking at the guest house that afternoon. In the aftermath of the execution, in fact, the legend would arise that he was completely shit-faced when he killed Jimmy Lee Gray.

After Bruce had gone off, Hocutt left the MSU himself and walked the 300 paces down the dirt road to his home. He hadn't slept well the night before and wanted to doze off, but he got up a minute after he lay down and went to the bedroom of his six-year-old stepson, Chris. The window looked out across a muddy field at the death house's smokestack. For years, Hocutt had been looking at that stack when he told Chris good-night stories. It looked so different that afternoon. So did the pink guard towers of the maximum security unit, which he had grown up loving. Everything looked different, far more vivid and real. Hocutt realized his heart was beating faster than usual, that it had been since he got up that morning. He was exhausted, and went back to his room and lay down, though he realized the moment his head hit the pillow he'd never get to sleep. He lay there, lost in thought, then looked at the clock. Four hours had passed since he'd walked back that afternoon.

He was at the death house in his officer's whites by 7 P.M. to sit out the final hours with Bruce, Jones, van Landingham, Childs,

Gregg, and Fulcher. Bruce had measured out a pound of cyanide from the plastic half-quart bottle van Landingham had given him after supper, and put the poison in a black metal locker in the witness room. Jimmy Lee Gray had started in on his Mexican food at a desk in the sergeant's office. Barring some change of heart at the Supreme Court, it would be the last time he ever stepped off C Tier.

Three hours later, the Supreme Court's 5–3 decision not to hear Jimmy Lee Gray's cruel-and-unusual plea one more time came down in a one-paragraph judgment: "This case illustrates a recent pattern of calculated efforts to frustrate valid judgments after painstaking judicial review over a number of years. At some point," Chief Justice Warren Burger concluded, "there must be finality."

The men assembled in front of the chamber. It was the first time in two decades anyone had been in there after dark, and the mosquitoes, which had made nests in the chamber room and chemical room, were going crazy. Hocutt had a case of Cokes brought in and opened the back door, then got a huge fan blowing across the doorway, hoping some ventilation would get the air and mosquitoes moving a little. There was nothing to be done about them, though. When they swarmed in the delta, you just sat, swatted, suffered, and made small talk. Hocutt was surprised at how much everyone was talking. You'd think people would shut up at such a time.

At 10:30, he noticed something unusual: everyone in the room had become "Mister." Except for Bruce and Jones, they had lived and worked together for years now, seven days a week. They hadn't called each other even by first names in months, but now it was, "How about another Coke, Mr. van Landingham?" "Still working on this one, Mr. Childs." The exception was Jimmy Lee Gray, who'd become "Jimmy." He'd been joined by a large group of Natchez church people, who had brought a pizza at Gray's request and were singing gospel and hymns. From the death house door, their singing sounded like the buzzing of mosquitoes.

By 11:00, the tension building in front of the fan had become deeply unpleasant. It was a strange feeling: of dissociation, though the men's attention was intensely focused on the task at hand; of impersonality, though everything from the nips of the bugs to the discomfort of the sweat staining their shirts felt extremely personal. It wasn't fear, though they all had the quickness to respond and the automatic deference that mark fearful people. It wasn't shame, but it weighed about the same.

A golden hearse from the Card Funeral Home of Indianola arrived at 11:30. The two doctors and a coroner were driven in through the MSU gate and around to the back of C Tier. The doctors were ushered in; the coroner was asked to wait outside. John Ledbetter, the sheriff from Pascagoula County, was brought into the death house by a man from the district attorney's office in Jackson. As the county sheriff, he was still mandated to "run" the execution. He'd do nothing but give the order to begin, but Hocutt felt Ledbetter belonged in that death house more than he did. The sheriff had been there when Daressa Jean Scales's body was fished out of the creek, when the news was broken to her parents, and when the Scales's marriage fell apart and they both left Pascagoula County. Jimmy Lee Gray's crime was the worst thing that had happened there for decades, and there were strong local feelings about the need to put him to death. Ledbetter was more than willing to share those feelings with reporters.

Somebody had to, because the media were there from all over the world, and no one at Parchman had any desire to talk to them. It had taken half of Hocutt's emergency staff just to see to the small city of press that had gathered outside the front gate: twenty-seven TV stations, each with a mobile home's worth of gear for remote hookups; a four-story antenna tower from a European satellite radio; a good fifty cars full of print people. Hocutt received calls every half hour, updating him about the traffic out there. His interest, however, was mostly with the golden hearse out back. It wasn't just the color, a bizarre one for a hearse. It was just that it *was* a hearse. There was no one dead here.

A few minutes after Ledbetter's entrance, Hocutt and Jones put on their gas masks, aprons, rubber boots and gloves and, at 11:57, got busy in the chemical room. Hocutt did the work, pouring a half dozen 500-milliliter bottles of distilled water into the sink, followed by the sulfuric acid. He looked down at the acid cooking, in huge, mysterious plops, then he and Jones locked stares through the fogged-over goggles of their gas masks. Jones looked down at the sink, clearly amazed at how big the plops kept getting. When the gurgling finally stopped, Hocutt sat down on a five-gallon bucket and took a breather. Jones was standing at the sink, still watching the acid, dripping sweat. He had forgotten everything he'd been told that afternoon, and in the small, close room he seemed disoriented and totally blank, like people get when they panic. Hocutt wanted to tell him not to worry, that nothing was going to happen, and no one was going to die.

With a sound of the last night cell's door opening, Hocutt and Jones stepped out into the doorway to the chamber room. Jimmy Lee Gray was being led by majors van Landingham and Childs to the chamber doorway. Chaplain Padgett, a step behind, placed a hand on Gray's shoulder as he entered the chamber with the two majors. Gray's face was ashen as they eased him into the chair, but he seemed more curious than frightened, watching the majors strapping his arms and legs, then his waist and chest. It occurred to Hocutt that Gray was the only man there who hadn't practiced the execution. Gray said something unintelligible under his breath to van Landingham, then looked at the chamber door as the two officers stepped out, leaving him alone in there. His eyes closed.

Hocutt was still waiting for the last-minute reprieve when a man from the district attorney's office gave an all-clear knock on the death house door, signifying there would be no further word from Jackson. The death warrant was read quickly, someone locked the death house door from outside, Sheriff Ledbetter said, "Let us begin," and the death house went silent. The natural impulse to stop a man's murder flashed through Hocutt's mind, then passed just as quickly.

Bruce stepped in and out of the chamber, taking only two or three seconds to leave the cyanide behind in the cone-shaped dispenser below the chair. He closed the door behind him, twisted the wheel, hit the exhaust fan, waited for the gauge to rise above 2, then without looking at Gray threw the lever down, putting his shoulder into the motion. He looked through the window by the lever for a second, to see that the cyanide had steeped in the sulfuric acid, then turned his back to the chamber and fired up a Lucky Strike.

The gas rose quickly, in a concentrated, small cloud. Hocutt and Jones still had their masks on and raised them at the same time that Gray took three deep breaths and moaned lightly. Hocutt realized he'd been holding his own breath and exhaled. Gray's head swayed from side to side, drunkenly, and fell back. A few seconds later, his head dropped forward easily, went upright and swayed, then dropped again. It stopped moving altogether and he appeared to lose consciousness. Less than a minute had passed.

Beyond Gray's slumped form, Hocutt saw the witnesses for the first time: the Reverend Padgett, head bowed in prayer, and Gray's lawyer, Dennis Balske, leaning against a wall in defeat, an elbow raised and his gaze at the floor. The witnesses were close: except for the three windows they sat behind, they almost seemed to be in there with Gray. Superintendent Lucas and Bobby McFadden, the investigator from the Highway Patrol acting as Lucas's escort for the evening, sat behind Gray's left shoulder, next to a man named Don Cabana, who was witnessing in his capacity as deputy warden at the state prison at Jefferson City, Missouri, one of the six states that still had a functioning gas chamber. The press witnesses, to the left, scribbled furiously. None of Gray's family was there, though he'd spoken to his mother, brother, and father earlier in the day. The families of Daressa Jean Scales and the Princes, parents of the sixteen-year-old Arizona girl Gray had murdered in 1968, had also chosen not to attend. Mrs. Prince, who had become director of rabies and animal control in Mohave County, Arizona, told a local reporter that morning that she had no desire to watch any man die, but that she was more than willing to go down to Mississippi and throw the lever herself on Jimmy Lee Gray: "I've put better dogs to sleep than that man," she said.

"God *damn!!*" Bruce suddenly screamed out, his back still to the chamber. Hocutt's heart stopped in his chest. It was something Bruce had probably meant to whisper, but it came out like a field holler. Hocutt turned his eyes from Bruce to the chamber and Gray's head, which was still bowed forward, as Hocutt imagined it should have been. He looked at his watch. More than ninety seconds had passed, and the gas was floating slowly and randomly across the chamber. Van Landingham, Gregg, Childs, and Ronnie Fulcher were staring gape-mouthed at Bruce. So were the doctors.

"God *damn*," he screamed again. "I told you she'd still work."

Hocutt and Van Landingham exchanged glances, then looked at Childs and back at Bruce, who was drawing deep from his cigarette, burned three-quarters of the way down with the entire ash still on. A huge belching noise came from inside the chamber, and the words, "Oh, Jesus, no," came to Hocutt's mouth as he looked away from Bruce to the chamber. Later, he wouldn't be able to remember if he actually said those words, or if his friend Ronnie Fulcher said, "He ain't dead, Donald" or just conveyed it with his eyes.

Gray's head was up and his eyes were wide open. A yellow foam sizzled from his mouth, and he was bucking up against the straps with what seemed a superhuman effort, then collapsing, almost shriveling, into the chair. Two long groans came from deep in his throat, followed by a horrible gasp, and Hocutt looked imploringly at the doctors standing to the right of the chamber. "He's dead," said the one running the EKG machine. Hocutt stepped past Bruce to check it himself. The green line seemed, if not flat, then awfully low. Gray's head fell again, then rose a second later as his shoulders squared up. His face was now contorted and red veins bulged in his neck as he stared directly ahead, roughly in Bruce's direction. Then he banged his head back against the pole behind the death chair. The noise it made was hollow, metallic, and sickening, and seemed to reverberate in the silence of the death house. Chaplain Padgett and Don Cabana stared past Gray from the witness room to the doctors and at Hocutt on the right side of the chamber, clearly wondering what they were going to do.

Gray's head banged back again, and Hocutt saw what at the

time looked like a flurry of activity. It was only two of the reporters, checking watches, and Balske, turning to Superintendent Lucas to say something. Through the glass of the chamber, he also heard one of the print reporters behind the left window ask, "Is this the way it's supposed to go?"

Gray's head slammed back again, then again a minute later, then twice in succession, hitting the metal pole so hard the second time the chamber shook. The foam had stopped sizzling in his mouth, but he still stared ahead at Bruce, who stared straight back. The stringers from AP and UPI, the reporters from the *Clarion-Ledger* and the *Daily News* in Jackson, and a half dozen people from the Pascagoula papers and TV station started counting how many times Gray's head smacked the pole. Each counted a different number. Hocutt felt like he was in a car wreck that wouldn't stop happening, with everything going in slow motion. On what was later agreed to be the eleventh banging of Gray's head, Superintendent Lucas turned to his escort Bobby McFadden and said, "Let's go."

A telephone in a white box at the entrance to the witness room rang as Lucas turned to leave. He picked it up, spoke a few words, then turned to the room. "The execution has been completed," he said. "Sentence has been carried out." Eight minutes had passed since the gas rose, but Gray's head had banged seconds before the phone had rung, and was still straining back toward the pole in a way that made him seem quite alive. His face was turned toward the doctors' window now, and his eyes were still open, though rolled far back in his head. Balske and the two wire stringers remained where they were, clearly defying Lucas's attempt to get them to leave.

A knock came on the witness room door, and Deputy Warden Joe Cooke, a former military man with an imposing air, came in with a large guard. "Gentlemen of the press," he said, "that was the pre-agreed-upon signal for witnesses to leave the room." They filed out quickly. On the other side of the witness room door one of the journalists asked, "He hasn't been pronounced dead, has he?"

"No questions," said Lucas.

•   •   •

Ten minutes later, Bruce tripped a second lever, opening the damper above the ceiling fan to release the vacuum. He followed with a plumbing of caustic soda and distilled water, then tripped a third lever that brought fresh air from the room into the chamber. Two gallon jugs of ammonia, which he'd placed to the left of the lever, were emptied in and their vapors were pulled up by the small fan, quickly detoxifying the chamber. When the beakers turned scarlet, less than five minutes later, Bruce put on his mask, gloves, and apron, unsealed the massive door, and went in to dust down Jimmy Lee Gray. It was a strange-looking procedure. Unbuckling straps with his old, practiced hands, he raised first the arms, then the legs, and patted each down. Then he rubbed the joints and bends of all four limbs with his rubber gloves, and wiped Gray's neck and face before moving up to the head, which he spent almost a minute on, getting his fingers deep in the hair, as though he were shampooing it. At his bidding, the two doctors went in to officially pronounce death. It was 12:47.

Hocutt opened the death house door to let the coroner into the chamber room to conduct his inquest. The man spoke briefly to the doctors, then prepared Gray's death certificate. It was ready by the time Bruce came out, and he and the five men of the execution team were sworn in as jurors, qualified as such to sign where the coroner had made X' s, agreeing that Jimmy Lee Gray had died of poisoning by hydrogen-cyanide gas on the morning of September 2, 1983, in a lawful execution ordered by the Supreme Court of the State of Mississippi. The cause of death had been changed from Homicide to Legal Homicide. Each man was given a five-dollar bill, for jury pay, then van Landingham, Fulcher, and Childs said good night. Bruce said that he and Jones would be leaving too. Hocutt showed them to the death house door, said, "Good night, Mr. Bruce," then turned to look at the chamber. He and Bobby Gregg were the only ones left in the death house, except for Jimmy Lee Gray, and Hocutt was halfway convinced he was still alive.

The two big men stepped in the oval doorway. Hocutt grabbed Gray's knees, Gregg got him under the shoulders, and they lifted him out of the black chair and through the doorway toward the last night cell. Inside, a metal table had been laid out, on which they were to strip him, hose him down, and reclothe him in a black suit the Pilgrim Rest parish had provided before taking him out to the golden hearse. The body was heavier than Hocutt would have imagined, and it took both men some effort, despite their size. Gregg got his half into the last night cell, but as Hocutt cleared the jamb, Gray's right leg and ankle suddenly clamped back on his hand, and Hocutt screamed.

"What, what, what, what, what?" yelled Gregg.

"He ain't dead, Bobby," said Hocutt, staring at Gray's hard and cold left calf muscle, gripping his hand. "Fucking guy ain't dead!"

Gregg's eyes widened. Then he saw that Gray's foot was caught in the jamb, bending back the bottom of the leg, and he started to laugh. It was a cold, ugly chuckle that Hocutt joined in on before he backed up to let the ankle through. Then they hoisted the body up on the metal table, stripped him, and got to work with the hose.

When Gray was unquestionably free of residue, five minutes later, the coroner came in to take blood, a disgusting process Hocutt could barely bring himself to watch: Between the needle and syringe, the aspirating tool was a foot long. The coroner looked nervous too, and kept his distance while he jabbed the needle in, like he was standing at a dart line in a bar. It went quickly into Gray's sternum—Hocutt couldn't tell if it had gone in the stomach, heart, or aorta, and wasn't going to ask. He hated needles.

When the coroner had left with a full syringe, Hocutt and Gregg began dressing Gray, which was easier on the nerves than Hocutt would have thought. When they'd gotten on the pants, socks, and shoes, Gregg nodded to Hocutt to look behind him. At first, he saw the clock in the chamber. It was past 2 A.M. Then he saw that the last five men still awake at the maximum security unit—two guards, a security sergeant, and two trustees, who had probably stayed around for the scraps from Jimmy Lee's dinner—had snuck back into the death house to watch them. They looked like the Marx Brothers, their heads stacked one on top of each other at the window to the last night cell.

Hocutt took Jimmy Lee's hand, which had gotten cold, and slid it slowly onto Bobby Gregg's wrist. "Yah-yah-yah-yah-yah," Gregg stammered. "Donald, goddamn it. Now don't go doing shit like that."

Birds were chirping when Hocutt awoke from a dreamless sleep at ten the next morning. Behind their song was a drone he hadn't heard coming out of his sleep in a long time: the sound of twin-engine planes, dusting crops. Their noise was constant in the delta this time of year, but it didn't usually greet his waking. This was the latest he'd slept in months, he realized, but his head ached regardless, with the kind of exhaustion only several good nights in a row would relieve. It was worse than a hangover or the guilty conscience that comes with it. It was more like the nervous enervation he'd feel on the third or fourth morning out on the emergency rescue team—butterflies you don't have the time or peace to deal with.

Along with the $255 he'd made last night, however, he had the day to himself, his first true day off in years. There was no need to be on call or even to be a family man. It was a Friday morning and his wife, Patty, was at work in the infirmary. His stepson, Chris, was out, doing whatever he'd be doing on a morning before school started for the fall term. It seemed strange that people were doing normal things. This morning, everything felt different to Hocutt. His mind, he told himself, was still geared to the protocol, still expecting every tick of the clock to be prescribed and followed mindlessly.

He decided to drive up to Clarksdale and get some things done that he'd been putting off for months. He'd bathed after coming home the night before and looked in on his family, but he took a long hot shower before plowing into the truck for the drive up 49W. It was a beastly day. Somehow, he'd thought the weather would have broken.

At the wheel, he was also surprised to find himself thinking about Jimmy Lee Gray—not as a convict or as a child murderer, just what he was like. In the past few months, Gray had begun asking about the

meaning of words and sentences in papers that his lawyers and the
state and federal courts had sent. Hocutt had been able to help with
some; others, like denial of *certoriari*, were harder, but they'd all been
pretty easy to intuit: one of these days, Jimmy Lee Gray was going to
be asphyxiated in the gas chamber for the murder of Daressa Jean
Scales. With each question Gray had asked, it had occurred to Hocutt
how straightforward it all was. In the words of William Boyd, the
special assistant attorney general who had fought Gray's constitutional appeals for four years, "Time had come to fish or cut bait."

But the death also gave Hocutt a fresh look on a huge chunk of
time that had passed since he'd first seen Gray in the bullpen, when
he was still a guard. Hocutt had since risen to sergeant, lieutenant,
and chief of the emergency rescue team, and he'd gotten married.
He hadn't liked the sight of Jimmy Lee Gray when he'd first seen
him in 1977, and as he thought about him on the drive up to
Clarksdale, freed from any judgment about his sentence or crime,
now that he was dead, he realized he hadn't liked the sight of him
any better as he stepped into the chamber.

Still, the actual punishment seemed strange. Ten miles up 49W,
the question came to him: Why was it hidden? If it was supposed to
help stop rapes and murders like that three-year-old girl's, why wasn't
what he did last night getting shown to people? If it was meant for the
vengeance of that little girl's family, and for that family in Arizona,
why weren't they there? And the chamber itself was awesome in
action, but he had to wonder: what was the point? He and some
guards could have taken Gray a few hundred yards into the woods,
strung him up, then stayed out in the pecan orchard with some beer
iced down and enjoyed the beauty of the day.

This morning, everything he saw on the road seemed full of life
and meaning. Despite his teenage years spent in Arkansas, Hocutt
was a man very much of the delta, defined by its agriculture and
machines. He looked at the various cotton pickers, hoes, and forks
on the drive up to Clarksdale as though they were telling his own
life: both the paths he had taken and the ones he'd given up for a
career in corrections. The combines, down from Tennessee for the

fall harvest, were on the roads and fields, so huge their headers needed to come down for even the highest overpasses. There are rivers and bridges everywhere in the middle delta, snaking out as the floodplain recedes from the Mississippi. The waters had made this place fecund enough to grow cotton easily, and where it no longer paid to grow cotton, the land got sculpted into ponds and levees for catfish farms. Morning was feeding time for the catfish, and from Tutwiler to Clarksdale, the tractors cut their slow, straight paths down the levees, trailing white feeder booms that sprayed the ponds with protein pellets. In their wake, the waters shook with the feeding frenzy; that late in the season, the ponds were packed as thick as a New York City subway car at rush hour, and the fish got to be three feet long, easily.

The moist red earth was being turned along the sides of the highway, an early start on the winter rice seeding. The weeds and insatiable grasses were soaking up the insecticides and herbicides dropped from turboprop Ag Cats and Cessnas. Even the slabs of asphalt and concrete jackhammered up along the sections of 49W that were getting repaved near Clarksdale were getting broken up with backhoe buckets, put through grinder trucks, crushed into thin slagheaps half a mile long on the shoulder, then plowed back under with earthmovers to make the foundation for the new road.

There were bodies of lynched men buried in these fields, meadows, and groves from one end of the delta to the other. None of them had gone to Heaven with the seven years of warning Jimmy Lee Gray had gotten, with a five-dollar coroner's jury convened by the State of Mississippi, or with some hideous sternal puncture. They certainly didn't get a burial like the one Jimmy Lee had gotten early this morning, though Hocutt knew a few of the spots where those people had gone into the ground. Some of the ones he didn't know were marked in the spring by patches of white tulips—a sure sign down here—or they were known by other people driving this road. Hocutt recognized a number of these people as they passed, and they exchanged nods. They would have been nodding differently if they'd known where he'd been eight hours earlier.

He'd be sure to tell them next time they met. There was nothing he enjoyed more than making people's eyes pop, and there were no secrets down here.

So why had Jimmy Lee Gray been done quietly? It made Hocutt feel a bit like a common murderer, or like a rat, hiding in the dark of a barn in the middle of the day. If he had acted on that millisecond's impulse to stop the execution, he'd have been known by name to every person in the state this morning.

He pulled into Clarksdale before noon, feeling ravenous. Outside the McDonald's, he got the *Clarion-Ledger* and the Memphis *Commercial Appeal* from the vending machines, then went inside, got a Quarter Pounder with Cheese, and sat down to read about the execution of Jimmy Lee Gray.

It had the front pages of both papers. Everyone in McDonald's was talking about it too: the girls behind the counter, the old folks, still working the bottomless cup of breakfast coffee, some truck drivers from Tupelo and Tennessee, in for an early lunch. None of them knew they were sitting next to the man who had done it. Hocutt unwrapped his burger slowly, listening in. Then he started to eat—slowly, but with more appetite than he'd had in weeks.

The *Clarion-Ledger* had laid a heavy hammer on the state for having botched the asphyxiation, and the eyewitness report, which was also done up as a column in the Tennessee paper, was clearly meant to be harrowing. The author, a Vietnam vet, wrote that the execution gave him flashbacks to the Mekong Delta, that he couldn't separate those two deltas in his mind now, but that what he witnessed last night was worse in its way than anything he'd seen in Vietnam. "Jimmy Lee Gray," he began, "died gasping and choking in the cyanide-filled gas chamber at Parchman prison early Friday for the murder of a 3-year-old girl who came to his apartment to play with his kittens."

A story on the same page reported that Morris Thigpen, the Mississippi corrections commissioner, had held a press conference after the execution, which had been attended by Bruce. There was even a quote from Bruce, who said he hadn't seen anything unusual about Jimmy Lee Gray's execution. "You people just hadn't ever seen one," he'd said. Thigpen admitted only that he was "disturbed by the feeling of some folks that we were trying to rush them out, or trying to hide something," then told the reporters that witnesses to future asphyxiations would be briefed by doctors in advance. They would also be allowed to remain longer than Jimmy Lee Gray's witnesses had. The doctor's official log, he added, had Gray dying of cardiac arrest two minutes after the gas dropped.

"That man was breathing when I was told to leave ten minutes later," Dennis Balske argued later in the same story. "And he was still alive." The *Daily News* reporter, Greg Kuhl, seemed to agree with Balske: "Gray appeared to be struggling for life when the witnesses were asked to leave."

But the people in McDonald's, talking about Jimmy Lee Gray's punishment over coffee and burgers, were clearly not troubled that he had suffered a painful-seeming death. In fact, they seemed happy about it. The state owed Jimmy Lee Gray nothing, least of all to take it easy on his punishment. Had it been easy for that three-year-old girl to choke to death in a mud bog?

Hocutt had worked at Parchman State Penitentiary since 1975 without a single vacation, on duty or on call every minute since making lieutenant four years earlier: on the tiers, on the grounds, in the fields, on the emergency rescue team, or working inmate teams ordered by Jackson to assist relief from tornadoes, hurricanes, and floods outside the penitentiary. He'd put in double shifts, triple shifts, and he'd risen with unprecedented speed through the ranks. But it was only now that he understood what it meant to be an instrument of the state—how it defined him, schooled him, and chastened him—and the understanding came with a solemn finality. To call what he had done twelve hours earlier "a job" was absurd. That had nothing to do with employment. To say it was "the law" either evaded the truth or missed the point altogether. What he had done was the right thing to do. And it wasn't some abstract will of the people that he'd carried out. It was the will of the people in that McDonald's.

CHAPTER FIVE

RANDOMNESS:
THE MAKING OF A MODERN EXECUTIONER

DONALD HOCUTT HAD BEEN LOOKING FOR A FEELING of that kind when he saw Parchman's classified for guard jobs in the Indianola *Enterprise-Tocsin* in 1975. He really didn't know what he wanted—a sense of belonging, of empowerment, of being necessary, of being in charge, of simply understanding where he was—but he was desperate for it regardless. At twenty, he was a mistrustful, ambitious kid who had spent his life swinging wildly between exhilaration and bitterness. Of late, he'd acted pretty much on the latter, and it clearly wasn't getting him anywhere. For four years, he hadn't spent more than eight months in any one place, and had only one friend, Ronnie Fulcher, who had started work at Parchman the year before. Hocutt had spent the year losing a full athletic scholarship at a small college in south-central Arkansas—"chasing pussy and whiskey instead of footballs," as Fulcher put it—and that was the best opportunity life was likely to bring. He had run through almost every entry-level job the delta had to offer, and was looking at the next forty years from the padded seat of a bulldozer.

He drove his canary-yellow Thunderbird up 49W to Parchman on Monday morning and filled out his application. The list of his jobs and salaries in the Previous Employment boxes looked strong as he wrote them down, but after a group interview four others were kept for further screening and he was sent home. Hocutt couldn't tell if he was too green or too hard-edged for the job, too dumb or too smart-ass. He certainly looked the part: The place was filled with six-foot, 250-pound, five-sandwich-eating country boys from poor, tiny local towns like Drew, Sledge, Marks, and Itta Bena. The group interview had looked like linebacker camp at a Southeastern Conference school.

He called the penitentiary three times over the next month, and finally got another interview, on the second Friday in August. The drill was the same. All he could think to do at the group interview this time was to make his desire to be there clearer than anyone else's, and that was easy. After the descriptions Fulcher had given him of the penitentiary, Hocutt wanted the job badly. An hour into it, he was called out in the hallway and told to come back on Monday at 8 A.M. After two weeks of classroom training—dos and don'ts on filing reports and restraining and counting inmates—he was sent off to the maximum security unit with another recruit. The conditions were extreme: Fridays off, $643 a month, and no overtime or danger pay, though the MSU was the most dangerous building in the State of Mississippi, and so understaffed it was clear he'd soon be putting in double shifts. It was a baptism by fire. The other kid lasted two weeks. Hocutt asked for the four-to-midnight shift, which he knew was the worst available. He told his supervisors the shift freed him for day classes at Delta State Community College, where he'd enrolled for the fall as a criminal justice major, but he knew his odds of making it at Delta State were no better than in Arkansas.

His plan was more realistic. As the saying went, they were offering him chicken shit at the MSU, and he was going to turn it into chicken salad. He'd done it with almost every job he'd had and he knew the trick. You took the dregs, then went back for seconds. He wasn't the only recruit with that idea, but his determination was greater. In classes, he'd seen the majors who ran the prison—silent, contemptuous-looking men with tall, Smokey the Bear–style mission hats and starched white shirts (everyone else wore blue). He would do anything to become one of them.

A horrible smell, half chemical, half institutional, hit Hocutt as he walked the MSU's length the first time, gawking down the four tiers of cells leading off the main hallway. A few days earlier, the men on B Tier had set fire to their mattresses, and the guards had just let them smolder. The morning sergeant told the incoming shift who had been acting up, but it was evident: Both A and B tiers were ankle deep in water from toilets that had been backed up, and the plaster ceilings were blackened with a plasticky grime from fires set to the recently installed Plexiglas windows. The noise level was high, which was perplexing: Hocutt couldn't actually see anyone. The only references for these faceless, screaming people were a series of arms and legs sticking out of cells and a big board outside the main office, which listed each inmate's name, crime, sentence, and county of conviction. Hocutt was surprised how many sentences still read *Death*—roughly one in six. What had been death row (C Tier, essentially) was near capacity, even though *Furman* had commuted those sentences from death to life imprisonment three years earlier.

One inmate was visible, but Hocutt didn't realize what he was at first: a "new man," who had arrived before the shift change, brought directly to the MSU by the county police. They had probably helped MSU guards move him into the bullpen, gotten releases signed, and taken off. The man stood there the first hour of Hocutt's shift, unchained and unwatched. Handcuffs didn't exist at the MSU—there was no money for them. Hocutt had no illusions about the prison's finances. In 1975, Mississippi didn't even have a department of corrections: Parchman's chief warden was called "superintendent" not out of nuance, but because he *was* corrections at a state level. Below him were two deputies and three majors, and each camp had a sergeant. These few men were the state's "DOC"; the "Class of '75" Hocutt was coming in on was the first multiple hire in years. The instructors in class had done nothing to hide the hardship, but the extent was staggering.

On a rare excursion outside class, Observation Day, Hocutt had gathered that work at Parchman would not necessarily happen according to the rules and procedures he had been taking notes on. A shouting match at a convict football game in a low-security camp had brought the point home. The game was awesome: no pads, few rules, full contact, and it stopped every third or fourth possession while someone limped or was carried off the field. They were good athletes too, better than some of the Division II players Hocutt had scrimmaged with in Arkansas. He sat on a wooden stoop of the camphouse and visualized offenses running out of an Oklahoma-style wishbone, and defenses in proper coverages, instead of the punch-and-run swarm they moved in.

The ball went over the camp fence and landed near an armed guard. He threw the ball back, but when it happened again, the guard just stared at the players, coldly, while they asked repeatedly for the ball. "Go home, asshole," one of the inmates finally yelled, admitting defeat. At first, Hocutt couldn't understand why they didn't hop the fence, get the ball, and keep the game going. Then he realized it would be jailbreak, and the guard could shoot. "Go home," the inmate taunted. "Someone's sticking a big black dick in your old lady."

"Not as big and black as the one you *will* be shitting on tonight," the guard said.

That was the end of the conversation. It was nothing but a game of "dozens," but the coldness on the guard's face gave Hocutt goose bumps. There had been headlines about correctional violence at Parchman two years earlier. The federal court in Greenville had awarded an inmate's lawsuit over the use of "Black Annie," a hard leather strap—roughly, a cross between a blackjack and a barber's strop—that had been wielded remorselessly. That decision, hailed as a landmark human-rights victory, was a first for

Mississippi. In the guard's glance, Hocutt saw that Black Annie was easily replaceable.

His firsthand lesson came less than an hour into his initial shift. "You might want to check this out," an older guard advised him. "We call it the Short Course on How to Win Friends and Influence People." In short: a few guards entered the bullpen with riot batons and beat the new man to the ground. Hocutt assumed there had been provocation, though he'd know otherwise as time went by. It wasn't the first beating he'd seen, but he was struck by the professionalism of this one.

The guard explained the reason for the "short course" once the new man was off to his cell. There was nothing in the MSU you really wanted to call human, he said. It was Parchman's brig, a place for extreme people for whom the rules—even the prison farm's extreme rules—were too lenient: compulsive arsonists, violent rapists, the murderously insane, escapees, anyone who attacked a guard. "The MSU," he said, "is where the hunter becomes the hunted." Sometimes, the quarry got made in the cells. If it happened on the way to or from a cell it was known as "intermediate correctional counseling."

The MSU had been put up in 1954, the year Hocutt was born. A high-security camp known simply as Unit 16, its alleged mandate had been to protect the prison's general population, but everyone knew better. It had been railroaded in from Jackson by the same centralist state senators who had put the traveling electric chair on the road in 1938. Much as the traveling chair had been a move on the regional power of Mississippi's eighty-two counties, Unit 16, which was to house Mississippi's death house, was an attempt to limit the power of Parchman's superintendent, a man named Marvin Wiggins. He had fought for years to keep executions out of Parchman, rallying behind him the locals who didn't want Sunflower County known as the place where people went to their deaths.

As the delta's main employer, Wiggins was one of Mississippi's most powerful men, for the simple reason that the state needed him more than he needed it. In 1954, Parchman was Mississippi's one penitentiary—the only other places for prisoners were two- to ten-cell jailhouses run by the counties or town sheriffs. More important, it ran at a profit. Parchman was a work farm, with well over 10,000 arable acres and a workforce that didn't stop until the crop was in—250 pounds of cotton per man per day. Wiggins fought the MSU, not only because he knew the motive of its inception, but because he knew his inmates would rape, murder, and hit the fences just to get off the long lines in the cotton fields and into the stay-at-home cells of the new unit. When the legislature voted to install the new gas chamber in the MSU later that year, Wiggins fought that even more bitterly. But the chamber went in the annex to C Tier regardless, and it effectively spelled the end of Wiggins's supremacy: until then, everything in Parchman had been his domain. Now, the most important thing that would happen at the prison—a man's death—would be the provenance of Jackson, with Parchman as its unwilling locale. Wiggins was the last warden to hold Parchman's superintendency longer than five years.

It was Parchman labor that built the MSU, however, and Wiggins made sure the state got the most remote and oppressive isolation building imaginable. It wanted secure prison cells; he made them six by ten, double-bunked, airless, and hot. In the summer the only thing that kept them from functioning as ovens were the small, barred windows at the top of each cell, and they let in more bugs than draft. In the punishment cells at the front of each tier—the "dark holes," which had neither windows nor bars—you probably could cook a man if you left him in long enough without water. There were few gates at Parchman, but Unit 16 was double-gated, fringed with rows of concertina wire. The fenced-in paddocks outside the cells, in which the state legislators had granted prisoners five hours of weekly exercise, were the narrowest kennel runs in the country. The only human concessions were the central hall connecting the tiers, a putative internal breezeway, and the pretty pink paint on the tower above the unit's front gate (it was the only paint available). No one bothered calling it Unit 16; few even bothered with MSU. Instead, it was known as Little Alcatraz. After the *Roots* TV miniseries, it became the Bodie Plantation.

It was, however, the farm's first "panoptic" building, and the first immediately recognizable as a prison facility. None of the other camps that had gone up in Parchman's fifty years had so much as fences around them. The prison's few gates served, as the maxim ran, to keep out curiosity seekers. This seemingly unsecure arrangement was made possible by the fact that Parchman was essentially self-policing. Trustees of long standing were armed from dawn to dusk with shotguns and 30.06s; known as "trustee shooters," these men were harsher and shot quicker than the prison's few employees, known as "free-world guards." At night, the inmates bedded down in "cages," cinder-block holding pens with one widemouth gate on the front of each. Inside each cage were twenty to a hundred men, overseen by a single free-world night guard who sat at a desk outside, generally a sadistic country boy with a cheap bottle on his desk and a cracked shotgun on his knee. Inside each cage, discipline was enforced by a convict goon known as a cage boss. Security out in the cotton fields was the trustee shooters' responsibility. If a trustee shooter lost a man, he lost his gun and went back in the cage, where the men he had guarded out in the fields would be waiting for him. It was a security system predicated on misery, and it worked.

The market for Parchman's cotton, however, was subject to the same pressures that hobbled other farms in the second half of the twentieth century, and by the early 1970s, civil rights suits dribbling in from state and federal courts in Greenville, Indianola, and Jackson made it impossible to work prisoners hard enough to turn a profit. Cotton production had slowed to a crawl by the time Hocutt started; by the end of the decade, when he was running work camps himself, the lawsuits had caused the prison to all but cease farming. The place had become a more classic penitentiary, and the MSU was its hellhole. There were no cage bosses on the MSU tiers (impossible in an isolation unit), so the guards raised a goon squad from their own ranks, and being part of that was the first step up the pecking order. Like other camps, the MSU had trustees to perform tedious chores, but as it was a closed unit, there were no trustee shooters. Only the free-world guards on the gun towers were armed: a rifle, shotgun, and pistol for each. Everyone else, including guards, surrendered their guns to the red basket that the guard in the pink tower lowered as they entered the unit. The trustees were brought in by van every morning to the MSU from other units; they worked in the kitchen and laundry, and rarely if ever saw the inside of the tiers.

There was no room for them. The population at the MSU had swelled to eighty men by 1975—not only rules breakers but anyone who wanted to "show his ass"—i.e., anyone with any fight left. Those types tended to stay on the tiers until the corrections committee decided it was time they come off. The corrections committee was usually slow to convene.

A man called Top Cat had landed in the MSU a few weeks before Hocutt started. He was only in for strong-armed robbery, but he was a genuine tough guy. He had fought the short course, even after two additional guards came into the bullpen, and he hadn't missed a chance to spit, throw waste, and yell obscenities at every guard or inmate he could reach or see. Within a week, he was in A Tier's "dark hole" for assaulting an officer. Rather than take his showers and hours outside in the kennels, he'd made a run at every guard who'd gone in for him, regardless of the beating he would get or the hellishness of the dark hole. It was the last week of August, and the temperature in there was well into the hundreds, twenty-four hours a day.

Hocutt's first order at the MSU was to join three other guards in taking Top Cat out to his kennel. "Don't dawdle," the sergeant said as he issued the four riot batons. Once on the tier, they had a minute to get him out, or leave him in. Then the gates closed. The dark hole was the first cell on each tier, just past the showers, so they wouldn't have to run the fecal gauntlet, but Hocutt could barely see the dark hole's solid-metal door, ten feet ahead of him. There was too much smoke, caused by fires set in the cells beyond. He'd walked through camps but had never been on a closed tier, and it was scary. His mind hardly registered the inmates' screaming, though they were yelling things that made him lose control. He wanted to be resolved when the dark hole door opened. If the man came out swinging, Hocutt wasn't going to back up, no matter what anyone else did.

Top Cat was at the mouth of the door as it opened, scarcely six feet, slender, with huge hands, and a bare chest covered with shiny sweat. His expression didn't change as he looked at the four guards, then his eyes settled on Hocutt. The two stared at one another, then Top Cat said "Fuck"—without a hint of surrender, though he took a half step back in as the door slammed in his face. His face was at the barred opening as the lock hammered shut, and he was screaming that they were just "bitch-ass, motherfucking, cracker-asshole redneck shit." Hocutt couldn't believe what he was hearing. When they had made it into the hallway and the double gates began to close, he looked back over his shoulder at Top Cat's vague profile, already vanishing in the smoke of the tier. He didn't feel enraged, as he normally would. He felt galvanized.

At dinnertime, the guards were issued fireman raincoats, boots, and hats from the office. Hocutt wheeled a trolley down B Tier with another guard, handing out meals and medications. Most inmates got at least two drugs, everything from cold remedies to antipsychotics. Prescription pills or vials stayed in a vault in the office, and had to be Scotch-taped to the number of the cell they were drawn for. The sergeant warned him to be careful not to mix them up. "Get a man walking on his hands and knees with the wrong antipsychotic," he said, "and you'll be down in Greenville, respondent to a king-size federal lawsuit." Almost every other inmate got some measure of Thorazine, but even the strongest doses didn't help. The men gulped it down like shots of whiskey, but from the first cell to the end of the tier, it was a toss-up whether feces, urine, or some kind of cocktail made out of shit and shampoo would come flying out when Hocutt put the trays in the waist-high slots.

When they'd finished serving the men, Hocutt looked back on B Tier, feeling like Lot's wife. The volume had dropped, but the noise seemed to echo in the smoke and water, as though he were in a canal. It made the hall seem timeless, at the same time that the collective fear and anger of the place gave it extreme presentness. In his temples, he could hear a pounding, as though adrenaline flowed in his veins instead of blood. It felt like chaos, and terrified him.

By the first head count, an hour later, he was getting used to the feeling. The MSU was smaller and sparer than he'd imagined, but there was a nonstop buzz to the building that was huge and new, and it corresponded to something inside him, something complex and very violent. Listening to the exchange about big black dicks across the fence, he'd realized that there was nothing going on here but incarceration and punishment, but he'd never thought the job would be this adversarial. Part of the exhilaration came in knowing he had the advantage over these people. They were behind bars, and he had the key. All told, they had murdered dozens of people, but that was outside these walls. In the MSU, they didn't even have the chance to surprise you around some corner, as they might in population or in the fields. They might make a run at you, get a book of matches, or draw a shank made from chicken bones, a door face, or a twisted-up Coke can, but Hocutt had access to a club that was two feet long, body armor, a bullet-proof vest, and a helmet if he needed it.

What unnerved him was that they kept fighting, regardless: all this hurling of vile things, as though there was nothing he or any other guard could do about it; all this insane screaming and name-calling, as though they would be on equal footing in whatever fight might break out. It made no sense, and the language rattled Hocutt. Even if these prisoners had some kind of moral or physical parity, the words would make him murderous enough to win any fight.

The exit at the unit's south end had heavy black bars, with spaces wide enough to rest a boot tip and his elbows when he leaned against the door to take his first breather, three hours into the shift. There was immediate release, leaning against the door, half his body out of the building. It made him feel like he was half in jail himself— he understood why the men spent the day with their hands and feet outside the bars—but it was a great place to watch the sun go down.

By ten o'clock, the lights around him were beautiful. Parchman was the thickest concentration of people from Memphis to Indianola, perhaps even Vicksburg, and its noise and brilliance, so anomalous in the profitless groves, flats, and wilds of the delta, made it seem like an oasis: the constant gunning of motors from the off-white Chevy Suburbans of the search and escort patrol, the free-roving officers who served as Parchman's cops; the crackle of the walkie-talkies and CB radios; the sounds of folks gathering on the porches along guard row at nightfall; an ambient buzz, from the screaming and clanging of other camps. Being in charge made it seem bigger. How did the phrase go? *Master of all I survey.* The penitentiary, with its laws and lawlessness, was a place of fantasy, and the MSU was its archest, most exotic part: so far off on its own, with the quasimilitary air of the cyclone fences and the Arabic quality of the pink guard towers. It was like a fiefdom, a pirate's cove, a caravansary, and a medieval dungeon rolled into one. Hocutt felt like he'd been there for weeks, even longer. Like himself, the MSU clearly had a lot more history than the twenty years of its existence.

Donald Hocutt had always been a history fan, particularly if there was a great man or feat involved, even more so if it had anything to do with violence, authority, or the military. Later, when he was in his forties and semiretired from Parchman, he'd call a psychic hotline number off the satellite TV to learn about previous lives. Told that he had been Ulysses S. Grant, Hocutt half believed it, though he would have much rather been General James Longstreet.

The middle delta he'd grown up in was the most backward area of the poorest state in the country, but the place was rich in history, and from an early age, Hocutt was fascinated by the Civil War—

the War of Northern Aggression, as they called it in his schools. His father, Marvin, got him a Union Army outfit for his sixth Christmas, and it made him a force in the Blue-Gray skirmishes the kids waged in the fields and cypress groves after school. Hocutt had the black kids on his side, which gave him numbers and staying power: their parents picked cotton until late in the evening. After his seventh Christmas, when he got a World War II–era machine gun on a tripod, his play was all German-American ambushes and sieges. That war had none of the Civil War's passion, but it had airplanes, and they became his fixation. The B-17 and Grumman Skyraider were the most beautiful objects imaginable, and he vowed he'd fly before he drove a car, even if it was just a single-engine plane.

He was a wide-open, high-strung kid, prone to brooding and full of ideas and secret plans. No matter how poor the Hocutts were, he kept himself armed and uniformed. He had Magumba rifles, cap guns, and BB guns, but from an early age preferred his father's real guns to his toys and fantasies. Uniforms, on the other hand, transformed him the moment he put them on. On Saturdays he rode with his father up to the county seat in Cleveland, and while Marvin window-shopped Caterpillars, he'd wander through the army surplus store, which was stocked to the rafters with World War II helmets, uniforms, and paraphernalia.

Before his twelfth birthday, the family moved to Garland County, Arkansas, where Marvin's two brothers were acquiring a big spread in the Red River Valley, thirty miles from the Texas border. According to their loan applications and the words on their pickups, they were in the excavation business, but in truth they were dirt farmers, getting land for clearing land. Eventually, they had 6,000 acres, plus a battery of earthmoving, paving, and farming equipment—trucks, tractors, bulldozers, draglines, backhoes, pickups, trailers, motorcycles, mobile homes, etc. It was hillbilly life, and utterly enjoyable. Hocutt worshiped his father, and his uncles were hardworking but relaxed people, practical jokers who had a great cumulative effect on him: he slowed down the notch he needed to, fell into his own, and grew up quick and large.

Marvin got colon cancer a year after they moved, and died in a few months. Though Hocutt couldn't have known it, he had just lived through the one year of his life he could genuinely call happy. By 1969, his mother had returned to Mississippi, and he was living alone in a trailer on his uncles' land. He continued to work for them, running the big machines and learning the mechanic's trade, and on Fridays and Saturdays he ran their moonshine and beer twenty-two miles down to Texarkana in his blue 1966 Mustang GT. But when he started high school in the fall, moving right into the varsity center and middle-linebacker positions, he spent less and less time on the farm, hanging instead with seniors, learning to drink vast amounts of Pearl and Schlitz without puking, and how to talk to girls. He had sexual charisma—as a newcomer, as a football star, as a live-alone fifteen-year-old—and it didn't matter that he lived in squalor. He was voted Wittiest Boy in Tenth Grade; the next year he won Most Popular. In his senior year, his full football scholarship came in from a local college.

He was essentially friendless, however, and increasingly aware, as he came of age, that he was something of a loner. He'd see his childhood friend Ronnie Fulcher on visits to his mother back in Mississippi, and the two would go out driving till 2 A.M. and get lost in conversations at a stop sign in the middle of nowhere. Fulcher would finally say, "Well, Donald, I guess this sign ain't going to change after all," and they'd go home. Fulcher was a year older, and everything Hocutt thought a man should be: educated, measured, and hard as nails. As time went by, Hocutt just found himself getting harder.

A few days after graduation, he wrote the college that he was taking a year off before starting his scholarship, said good-bye to his uncles, and drove to Tallahassee, where he'd heard he could pick up work underage, driving a bulldozer in front of the crews building I-10. The work was repetitive and the pay was dreadful, but he had his own room in nearby Montracello, at a $14-per-week flophouse called the Cherry Arms, and he thought he'd arrived. In fact, he'd become an alcoholic. By the fall, when he was in Fort Worth, driving a scraper truck at the new airport, he'd already realized how quickly the life of a hard-working, hard-drinking man got old. He started thinking about his father, and how wrong things had gone since he had died. Leaving work in the middle of a shift at Dallas–Fort Worth, he drove to the marine recruiting center in Texarkana and signed up. Barely seventeen, he lied about his age, but he didn't see why that would make a difference. He really wanted to kill someone. He was on the bus for Parris Island when the shore patrol came aboard and put him and his suitcase out on the sidewalk. By the end of the month, he was back with his mother in Indianola. He started college in the fall.

The following summer, when he saw the ad for Parchman guards in the *Enterprise-Tocsin*, there was no doubt in his mind that his last chance for respectability had come. Looking back, decades later, he guessed he was looking for a father figure or something of the kind at Parchman. He was also looking to kick some ass.

Hocutt's first encounter with near-lethal violence at Parchman came six weeks after he started at the MSU. Twelve men on B Tier had rioted the previous night and into the morning after getting plastered on prison buck made of yeast and sugar. They had all nodded off by noon and were being separated—three officers moving teams of two at a time—when Hocutt's afternoon shift started. Some went to the dark holes, a few to the hospital, some just to different cells or tiers. It was a difficult logistical task, deciding how to pair them off so they wouldn't rape or kill each other when they woke up and saw who they were living with, particularly with their heads aching with the kind of hangovers that bad, dehydrating sugar-whiskey brings. Having that many men in transit was a nightmare.

Two of the inmates refused to cooperate. They hadn't gotten any liquor, hadn't rioted, and were incensed at being separated just to make the "asshole pairings" work out. "Fuck you," one said when the team went in to get them. The one who said it, Pilot, was a tall, thin twenty-three-year-old who had made no problems at the MSU.

His older and stronger cellmate, Brick, was belligerent, and Hocutt knew right away Brick was the one putting up the fight.

He and two other guards were issued helmets with shields and bullet-proof vests and walked down to B Tier. Inmates from the other tiers saw them passing and started hollering that the goons were coming. When they appeared at the double gates of B Tier, the whole side of the unit erupted with threats and screaming. The other two guards had taken up riot batons when they'd dressed for battle at the office, but Hocutt decided he would be better off bare-handed. The doors to the cells were wide enough for only one man at a time. Hocutt, already well known as an MSU bad boy, would be the first one going in. Between the double bunks, the commode, and Brick and Pilot, there would be no room to swing a stick, which meant its only use would be theirs, if they could get it from him.

They were kicking and swinging the moment the cell opened, as expected—particularly Pilot; as Hocutt had learned, the lesser of two instigators always rattled the cage harder. He paid Pilot no mind, at first; it was Brick he wanted. The man saw it, and started backing up, first to the wall under the tiny window, then behind his bed, still kicking and hollering. Pilot hit Hocutt with a good shot behind the head, and Hocutt turned, picked him up by the scruff of the neck and groin, and threw him out of the cell, where the other two got to work with the batons. When he went back for Brick, he was surprised how easy it was to turn him, get an arm around his neck, and move him to the hall.

On their first lock-step together outside the cell, however, Brick broke free, turned, and squared off. It was token resistance—Hocutt had him around the neck again in no time—but the tier went nuts, and urine and fiery paper started flying from all directions. The fact that the man was putting on a show at his expense made Hocutt livid, and he started tightening up on the neck, until he could feel things inside it begin to move in a way they weren't meant to. *One more yank*, he thought. *Or just sit down*. With the angle and leverage he had, that would probably crush the carotid artery and kill the man. "You can turn me loose now," Brick muttered. "I've had enough."

Pilot was carried off the tier by the two other guards; Hocutt took Brick's arm, twisted it behind his back, and pushed him out of the tier. En route to the bullpen, and a brush-up on the short course that both men knew was coming, Brick turned to Hocutt and said, "I'm sorry, Officer." Hocutt was shocked.

"We have to live with these people," he continued. "It wouldn't have looked good if we just went along." Brick's tone convinced Hocutt, but he was befuddled: what concern could these people have for how they *looked*? Brick had come within two seconds of dying—for appearances' sake?

It took *Gregg*, and the arrival of the condemned, for Hocutt to work out the syllogism of fearlessness he kept running into at the MSU. The condemned were almost preternaturally docile. When they spoke up, it was usually for something the guards were happy to grant. They had calendars, but needed help doing the sums of how much time they had been on the row, or how much they had left. They read things in magazines that led to questions that reminded Hocutt of how stripped their lives had become. Who was the vice president? What is HIV? Did the Supreme Court ever come to Mississippi? Much later, in 1986, one of them had been amazed to read about the space shuttle, and couldn't quite understand Hocutt's explanation of what it was like to go to the moon. "It's been eight years since I saw the moon," the man said.

No matter how far the other inmates in the MSU had fallen, they had hope. With that hope came fear of what might still get lost, and from that fear came this indomitable anger, born of desperation. Watching the beaten manner of the condemned as they began to arrive in 1977, Hocutt understood the difference between a criminal and a condemned man. It was the most important trade secret a jailer could learn: If a man has any hope, you have to leave him a way out. Two decades later, he would have to learn to apply the lesson to himself.

Shortly after Hocutt started, he took an apartment twenty miles away in Cleveland with Ronnie Fulcher, with the idea they could just roll out of bed and go to school. Hocutt's interest in 9 A.M. classes, however, paled with each graveyard shift he put in and each fifth of Smirnoff he drank. He went through a bottle a night, and after the vodka there were redneck bars where he could drink moonshine or rotgut for next to nothing, shoot craps, and get into just about any kind of fight: old-fashioned slugfests with a bottle broken over the bar, shootouts with .22s—kept low, of course.

Hocutt left a roadside tavern in Cleveland late one night with a half bottle of vodka left. By the time he pulled into a taxi stand next to a fried chicken joint in the black part of town, there were only two inches left in the bottom, and Hocutt was so drunk he thought he was still in uniform and armed.

Two men sat at the counter, eating chicken. "Move over," Hocutt told them. "And get me a sack of chicken and be quick about it." Instead, one of them punched his face, then the two grabbed him by the elbows and a leg, and threw him out. Hocutt landed face-down in the parking lot and went back to the Mustang for his pistol. He had forgotten he'd put it in the trunk instead of the glove compartment, and wound up with the vodka. He downed the dregs, went back in the chicken joint with the bottle, and used it to smash the face of the man who had punched him. The guy went down, and Hocutt took a roundhouse swing at the friend and missed badly. A stool went over his head, and when he looked up he was back out in the lot, with both guys standing over him, stomping his ribs and face. When they started on his groin, he managed to roll under the Mustang. They threw gravel and stones at him, gave up, and went back in.

Hocutt got out from under the Mustang and into the trunk for his .38 Smith & Wesson service revolver, then crashed back through the door. He had known all along he didn't belong there, was becoming aware that one of his ribs had been broken, and he also knew he'd gotten his ass kicked fair and square, but he had to redeem himself. "Time to go," he yelled, firing two shots through the ceiling. Within seconds, he had the place to himself. He sat at the counter and said, "What's a white man gotta do to get some fried chicken around here?"

He was at the wheel outside a car wash half an hour later, eating the last of the chicken, when he saw blue lights in his rearview mirror. The cop who came out was a black guy Hocutt knew fairly well. He shined a light in the car and was shocked to see that the armed felon he had gotten a call about was Hocutt, grinning through a busted lip already twice its normal size, hunched over to nurse the cracked rib, his face and neck spattered with blood and sheened with chicken grease. "Jesus Christ, Donald, what're you doing over here?"

Hocutt smiled and finished eating his chicken. The cop asked if he needed help. Again, Hocutt didn't say anything. Through his windshield he could see the sky was beginning to lighten: no chance of getting to sleep and making class in less than four hours. It didn't matter. From the way the cop was looking at him—not as a white man, a drunk, or an armed felon but as a colleague—Hocutt knew his days at Delta State were over. He was a corrections professional, not a corrections student. He didn't belong in Cleveland any more than he had belonged in Indianola or Garland County. There was only one place in the world for people like him. He hadn't been born to it, like the second- and third-generation people at the farm, but he was Parchman to the bone.

Hocutt shared his first house inside the prison's walls with the MSU's sergeant, Roger van Landingham, whose family had a long history at the penitentiary. The sergeant was five years older, but he and Hocutt shared a wild streak. A year later, when Hocutt moved to a trailer house in the K-Nine unit, he was a Parchman legend, and the walls and ceilings of the old clapboard house he shared with van Landingham were riddled with bullet holes.

From van Landingham, who was more smart than tough, Hocutt learned that moving ahead at the MSU wasn't all about ruling with an iron fist, as he had first thought. There were some real sadists among the guards, whose violence he couldn't hope or care to match, but they didn't amount to much. Most of his colleagues ran inmates with prison commonplaces—snitches, favorite trustees, threats, and rewards—but those tricks discounted themselves as quickly as any cliché. The MSU was no different than high school. You did things the way they came, naturally, one at a time, kept your aura clean but mysterious, and people made way as you passed.

Being natural in a prison was a lot like it had been in high school. It meant knowing your place in the hierarchy: every order Hocutt received was gold-plated, and his only responses, other than "right away," came if he didn't understand, or if it would be appreciated if he pointed out something that might have been overlooked. Keeping his aura clean meant being known, half the time, as a crazy country boy. When inmates set fire to rolls of toilet paper in their cells, Hocutt, who was required by law to douse the fires, attached the water hose to the hot-water spigot instead of the cold. There were no more toilet-paper fires on his shift. The other half of his authority came from being a mystery man, and that was roughly divided between extreme moodiness and practical jokes, both of which fit in seamlessly with the fabric of prison life.

He had a knack for hanging a sign on people he didn't like and for getting into their world and making it miserable. When a guy on the row named Shine got caught masturbating to a Bugs Bunny cartoon featuring a female rabbit, Hocutt made so sure the word got around that Saturday mornings became hell for Shine. From the moment the Looney Tunes theme came on, the taunt of "Shine's fixing to come" would echo down the tiers like a yodel. Late one night, he got the executioner's gas mask out of the death house chemical room, strapped a lit flashlight into it, stuck it on a broom handle, and carried it to the window of a jittery murderer on A Tier named Percy. Hocutt eased the lit mask up through the barred window above the bed and twisted it around, moaning a ghostly "Per-r-r-cy, Per-r-r-cy," until the man woke up and started screaming. Hocutt had underestimated the rise he would get. Percy went off every night about the giant grasshopper ghost in his cell, and the men on A Tier complained so bitterly he had to be moved to a different tier.

Another murderer on A Tier, Dan, got drunk on buck one Sunday, when he knew that visitors would be in the MSU and he wouldn't land in the bullpen for the short course. Hocutt and another guard took Dan down C Tier past the last night cell and into the death house, opened the gas chamber's big submarine door, strapped him in, and agreed to forget about him until the sun went down. He was sober when they went back for him but shook and mumbled so much he had to be carried all the way back to the tier.

Pranks like that could land guards in federal court even quicker than a physical abuse complaint. From the first, Hocutt saw his name inside the perforated boxes on far more than his share of state and federal court documents: _____, Complainant, v. *Donald Hocutt*, Respondent. At times, the complaint was from his riding someone else's heat: taking blame for some excessive use of force, for example, if a senior officer or shift partner would have suffered in the docket more than he did. At other times, he pointed the finger where blame belonged. Each day brought its own judgment calls. If he didn't care for a guard, he might take a rap once, but he would let the man know it was going to be him in the docket next time. He despised officers who claimed an inmate "slipped" and fell on a bar during a handcuffing, and he was embarrassed when a guard tried to claim a man had caused a bump to appear on his own head. If you hit a man over the head with a stick, you should be man enough to admit you did it. You really didn't have to lie about using force, like some politician with his pants at his ankles. The job was political enough—as much as any crew he had been on, or any office in the world, he imagined. The difference here was that there was no quitting time.

Hocutt learned how to deal with the twenty-four-hour nature of his job from the old-timers, the ones whose people had been here for a generation or two. He got close quickly to a Parchmanite family called the Parkers. The brothers, Mike and Barry, worked at the MSU—Barry was on his way to becoming the unit's chief of security and a member of the execution team. Mike eventually ran the arsenal, where the cyanide was kept. Their sister, Sarah, ran the radio dispatch, which was the farm's central nervous system. Anyone moving prisoners had to call her, and she effectively kept a running body count for the prison.

A distant cousin of Hocutt's, also working at Parchman, married the daughter of Sergeant Pat Mooney, a soft-spoken, chain-smoking character who had been the first to run the MSU. Hocutt would bring a bottle to Mooney's rickety, unpainted rocking-chair porch and soak up the history. It was the first picture he got of the death penalty, and he was surprised how matter-of-fact it had all been: several times a year, the Mississippi State Supreme Court said *Kill this man*, the men met behind the MSU at midnight, killed him, then went back to where they came from. You would hear singing from the row, if the guy hadn't been there long enough to antagonize everyone, and occasionally you would hear about a problem with the lever or the mixture, or the condemned man putting up a fight or something. There were rumors about Thomas Berry Bruce showing up drunk: even that he had been too drunk once to carry out an execution. Mooney said it was just gossip. He had been around for every execution, and he didn't remember anything like that. More to the point, there was no one else who knew how to work the chamber. It had more moving parts than a four-stroke engine, and any slip-up could mean the operator and witnesses died, too.

Hocutt listened intently and fancifully to Mooney's tales, but it was hard to make much sense of the bigger picture. It was only two decades later, piecing together the farm history he had learned on that porch, that he saw the leap that had been going on when he'd started in the mid-1970s: a wave of liberalism and rehabilitation-minded bureaucracy that made him something of an anachronism even as he was hitting his stride. In 1976, on an assignment outside the MSU, he had been told to run a "gun line" one of the majors had set up on the prison's baseball fields, where a campful of men had been moved after a riot. A gun line was a southern tradition at least as old as the Confederate prison at Andersonville. You drew a line in the dirt and shot anyone who crossed it. The gun lines here were the first- and third-base lines.

One inmate thought that was ridiculous and walked up to the base path and nudged his toe at it. "You gonna shoot me?" he taunted Hocutt. "Look, I'm on the line. I'm moving over." Hocutt got a bullet in the chamber and raised the gun. "You gonna shoot me?" the man asked again. His tone was insulting.

"I honestly don't know," Hocutt said, aiming at the man's head. "Why don't you step over and we'll find out." That was enough to end the episode. Looking back, however, he could feel the movement into the "touchy-feely" age right on that diamond: two years earlier, no prisoner in his right mind would have stood up to a loaded gun like that. Two years later, even setting up such a line would have drawn a mammoth lawsuit in federal court: In late 1976, Mississippi finally established a department of corrections. Its rule book, coupled with the state and federal decrees arising out of inmate lawsuits, put an end to things like gun lines. It brought some pluses, such as schools run by the FBI or criminal justice departments of state universities like Florida, Texas, and Michigan. Hocutt took every class they ran—management, forensics, bomb making, firearms, clearing houses, hostage situations, and hand-to-hand combat. Classrooms, teachers, and tests had given him problems outside, but on the farm they just seemed to come easy.

He made sergeant the next year, at twenty-three, the youngest in Parchman history. There had been some deflation in rankings with the new DOC—now, there were lieutenants, captains, and even colonels—but the promotion was an unspeakable honor; Hocutt was determined it would be the first of many firsts. In September, he became the youngest to make it to the emergency rescue team; two days later, he put on the black camouflage ERT jumpsuit and went out with a rifle and the dogs of K-Nine unit after two men. It took almost forty-eight hours, but they found them, still on the grounds. Three other inmates were hitting the fences as they drove them back into camp, and Hocutt smiled. This was going to keep him busy. He'd loved the chase, which brought a $75 danger-pay bonus, and the black camouflage uniform had power; three days later, he still could almost feel it on him.

Hocutt joined the Freemasons, taking the lifelong vows of secrecy that many at Parchman did, if usually at a much later point in
life. On a Saturday night in the spring, he took another lifelong vow
that took some getting out of. He had been up in Clarksdale, getting "knee-walking drunk," and "woke up married," to a nurse
he'd met in a bar. She couldn't stand prison life, he couldn't stand
her, and, in truth, he had his eyes on another nurse, a wild, tough,
and beautiful country girl from Sledge named Patty, who worked in
the Parchman infirmary. She was getting out of a marriage herself
and had a little boy, Chris; to his surprise, Hocutt welcomed even
that responsibility. They were firing Parchman's wardens almost
yearly—Hocutt was already on his third—and he had pretty much
given up on father figures. He'd divined that the only way out of
the bitterness of losing Marvin was to become someone else's
father. Chris was a sweet little kid, and even looked a bit like him,
with about 90 percent of the air taken out. The only thing lacking
was the chance to court Patty: the MSU didn't leave much time.
Before the year was over, though, he was out of the unit, driving a
Chevy Suburban on the search and escort patrol, with the whole
penitentiary as his office. There was no doubt that he had arrived.

He and Patty were married, and by 1980 lived in a white house
at the end of guard row, 300 yards across an unused field from the
MSU—Hocutt's fourth house in as many years inside the walls. A
new warden, Eddie Lucas, made him lieutenant. He had little
choice. Hocutt was one of a handful of guys he could go to when
something had to get done. Death row had been building steadily,
and another maximum security unit was put up, Camp 32, a huge
facility. Hocutt was taken off the search and escort patrol and given
what was now called the "old MSU," as well as five other camps:
a trustee camp, a work camp, the women's unit (28), the building
they put homosexual and psychiatric cases in, and Camp 24, a
high- to medium-security camp that became the bane of his life.

Making sergeant had been an honor. The lieutenancy, however,
brought stresses that at times made life almost unbearable. It took
a few months to realize it, but the promotion spelled the end of a

phase of his life that, while having its ups and downs, was manageable. It had been one thing to get up every morning, put on a stick and a can of Mace, step inside a fence, and go to work. Now, every problem at the prison came to his desk, and it made days longer than they had ever been. The weeks blurred into each other, and life became chaotic. Hocutt had learned early to take it one step at a time; it kept him from getting overwhelmed. Now, there was nothing standing out from the stress to break life up into segments.

He had always looked forward to the Fall Rodeo, for example, particularly the Brahma bull rides, where officers would put a $100 bill in a tobacco sack, tie it between the bull's horns, and watch two or three dozen inmates go after it. As lieutenant, though, he was the one that had to get up early on the Monday morning after the rodeo, race to the grounds in his jeep, empty his gun into the dust to break up the knife fight that broke out between the men who didn't want any part of cleaning out the shit, then go up to Greenville two weeks later to explain why he had unholstered his weapon. Three days later, he was in court again, explaining his violence in breaking up a knife fight that had broken out in Unit 32.

With half of those suits, he had to stay overnight at the Ramada in Greenville, which meant losing the prime days of the fall, which meant losing football. When an entire week in October came up with no court dates imminent, Hocutt went down to Camp 12, which had the best athletes, and talked his way into becoming the farm's first free-world football coach. As Hocutt had envisioned on Observation Day, Camp 12 was invincible once he'd taught them the wishbone, a proper 4-3-4 defense, got a playbook half an inch thick printed up, and worked out a relay of wide receivers to send in plays. During practice, however, he was the one who had to respond when they threw the ball over the fence, just to make him go get it.

They never seemed to tire of fucking with him, and if it wasn't the inmates, it was the guards. He called his management style a "mild dictatorship." Anyone with a problem should feel free to come to him, as long as they brought the solution in with them.

They never did, and as time went by, he just got more sarcastic and impatient, and the lawsuits kept flowing. An inmate whose ear infection medicine wasn't doing the job wobbled in and complained his ear was ringing. Hocutt told him: "Answer it." It sounded funny to him, but two weeks later, he was up in Greenville, explaining what he meant by that.

After four years inside the walls, he needed out. Officers lived on the farm, however, and Hocutt wouldn't break rank. He tried Delta State again, but classes made him dreamier than ever. The Cleveland Airfield was a mile past campus. On his worst days as a rookie guard, sitting in class wondering what Cesare Beccaria or Jeremy Bentham had to do with dodging urine, he'd thought about walking in and asking what it took to fly a plane. This time, he walked out of a criminal theory class that was dragging, drove the extra mile, bought the FAA workbook, and brought it back to work and memorized it. Two weeks later, he aced the written test. After seven hours of solo, he became the youngest man at Cleveland ever to get an FAA pilot license. As per ritual, they cut his shirttail and hung it up on the wall with all the shirttails of the Ag pilots—as the crop dusters were known—and Hocutt walked onto the field a full-fledged stud-puppy.

He rented a yellow Citabra stunt plane and took off to the northeast, gaining altitude over the soybean fields and catfish farms and the silver rivers toward Parchman. It was a breathtaking sight: the partitions and boundaries of the property lines and the natural slopes created by the delta's recession from the Mississippi. To the naked eye, it was just varicolored rectangles, squares, circles, and oblongs, but from on high, if you knew the agriculture, you could see the non-Euclidean geometry that shaped the land; where the bedrock rose too high to let the water percolate, and the land had to be formed; where the soil densities and the grades and shapes of the aquifer made soybeans impossible, and the catfish farms began to dominate the landscape, west of Greenwood; where the mini-depressions of the tributaries settled into rice fields; and where the delta finally played out in a jagged, oyster-edge line and gave way to the massive emerald fallowness where kudzu swallowed up everything from fields to forests to riverbanks and the promontories of the highways and country roads.

The airspace was busy almost everywhere from 500 to 2,000 feet, though, and that disappointed Hocutt. He'd had fantasies about quitting the prison, going into business as an Ag-pilot, and finally getting some solitude, glamour, and real money. Up with the crowds of these guys, he knew it wasn't the romance he had bargained for—just another cutthroat business with refinancings, payment deadlines, and fractional profit margins. The blue sky and wispy white clouds got dull quick, too. What made flying interesting wasn't what was up here.

He turned 180 degrees and headed southeast for Greenville. The clearance below the Mississippi Bridge was as low as one could reasonably hope to challenge. He brought the plane down until he could smell the brackish water, got himself centered, went under the bridge with room to spare, then took off straight up the Mississippi River at full throttle, letting the barges, trawlers, and fishing boats know he was there. Hunting season had started, and when he buzzed the treetops the whitetail scurried from the deep privacy of the thickets. He was impressed by the vantage the altitude gave him, even in such densely wooded areas. Two years later, when he became the chase coordinator of the emergency rescue team, he requisitioned planes for difficult chases.

The deck crew on one of the big barges seemed to think he was making too much of a racket, and they shot him the finger. Even the oilers, blackened with grease, had come up from below decks to flip him. Hocutt dropped the nose a full forty-five degrees and gunned it, like a kamikaze pilot, and all hands hit the planks. Then he saw the cabin of the boat the barge was pulling, and yanked the stick back. He cleared the roof by inches, howling, climbed back up, hung a right, and headed off to buzz the control towers and concertina wire of the old MSU.

Planes were rarely seen on prison grounds, other than at crop-dusting time, and the roar of the Citabra above the death house smokestack woke the place up. Hocutt couldn't shake the sense of how drab it all looked, though. He couldn't see the building's pink paint from above, and the old MSU was just an ugly red-brick building standing in the middle of prime cotton-growing land. The smokestack looked like a relic from the Industrial Revolution.

He used a long pebbly patch of an old butter bean field on Unit 27 as a landing strip, and the majors and trustees ran up to check out the plane and yell at him for flying so low. The trustees got to work with soapy water and wax, and van Landingham, looking at the flight panel like he was memorizing it, told the young lieutenant what an immature asshole he was, just like Hocutt knew he would. Childs said he could actually see the screwholes of the big rivets on the bottom of the plane as he'd buzzed him. That was about as close to perfect as anything Hocutt had ever heard.

Then he saw where he was—in a butter bean field that some Jackson bureaucrat had decreed should go fallow, looking at some grinning snitches with stripes on their jeans, waxing someone else's airplane. Short of the major's bars, he had everything he'd set his mind on, but he felt hollower than when he'd started. If anything, the joyride made him feel more trapped—inside himself, inside prison—and further from whatever it was he just couldn't get at.

He'd finally get it, three years later, sitting in the Clarksdale McDonald's after Jimmy Lee Gray's asphyxiation. The feeling didn't last, but at least he saw exactly what he was looking for—a sense of being instrumental. Achieving it made him feel less alone, more public, more out there, despite the secrecy that had veiled the execution. There were a few on the execution team who sucked that secrecy up like mosquitoes. They were so good at keeping the secret of their involvement that there wasn't a person in the penitentiary who hadn't been sworn to keep it. They were Hocutt's friends, but their grandiosity filled him with the same contempt he felt for guards lying about their use of force when they wrote up rules violation reports. Killing a man was monumental enough. There was no need to make it bigger. The only one who got bigger for it was Jimmy Lee Gray, and he was just a psychotic with an inferiority complex so big a three-year-old girl had to die. He was a displaced

California kickout who had left an Arizona prison cell he should have died in, and wound up sitting in an oversize chair inside a machine in Mississippi that probably belonged in a museum more than a prison.

The randomness of it all impressed Hocutt more than anything. Jimmy Lee Gray wasn't born to die in that chair any more than he had been born to stand at the chemical room door, watching him die, or working in a prison for that matter. Hocutt rarely looked at life from on high—dodging urine and feces for years on end didn't encourage a broad perspective. When he did, whether it was down the nose of a stunt plane terrorizing the delta or with the second-by-second hyperawareness that comes with standing in such a doorway, the randomness was what he saw, playing out like a white cloud or the dance of a dingy yellow gas in a vacuum-sealed chamber. It was sometimes dreadful, sometimes exhilarating, and almost always mysterious. Life could go this way or that, but finally it went the way it had to: Thomas Berry Bruce had a deputy whose wife got sick one night in September 1983. Jimmy Lee Gray died in a gas chamber some politicians in the 1950s decided to put in a cotton field instead of in their own backyard. An engineer in Denver in 1938, or Salt Lake City in 1954, put a pole an inch behind a chair when he probably should have put it two inches further. And that difference gave Gray's death an impact that would change the way certain things happened in the delta. It would save the lives of a dozen condemned men who wouldn't otherwise have survived the 1980s. It got some people elected, some people fired, and it made Donald Hocutt a part of history.

CHAPTER SIX
AS FATE WOULD HAVE IT:
THE MAKING OF A RELUCTANT EXECUTIONER

OCTOBER 1, 1984. Thirteen months after Jimmy Lee Gray's death, the Parchman superintendency was given to Donald Cabana, deputy warden of the Missouri State Prison at Jefferson City. Hocutt took a rare immediate liking to Cabana, and saw his arrival as the last and best cog of a huge wheel that turned in sync with that execution. The wardency was a political position, and there was no Parchman event more political than an execution. A week before the asphyxiation, Bill Allain, who had been the state attorney general during Gray's last four years on the row, had won a photo-finish runoff for the Democratic gubernatorial primary against a strong female candidate, Evelyn Grady. It had been a classic law-and-order political victory, with litanies of Gray's crime, the state and federal denials of stay, and the gearing up at Parchman's death house dominating the news during the runoff. Evelyn Grady couldn't compete with the sound bites Jimmy Lee Gray's long-awaited date with the gas chamber provided her ex-prosecutor opponent.

Allain's election in November, and the subsequent changes in the Department of Corrections that led to Cabana's appointment, would likely have happened without the asphyxiation going as it had. It was time for the wheel to turn, and it came up on Hocutt's lucky number. Within months of Cabana's arrival he became Major Donald Hocutt—Cabana didn't even bother with captain—and Hocutt's mid-management nightmare was over. He had prospered under Eddie Lucas, who'd had no qualms about letting Hocutt take on all the shit-work he wanted. But it had stayed at that level, and it galled Hocutt to watch the advance of fellow officers with less to offer but more political savvy. Lucas was a politician himself, a master of the golden rule of his office—keep your head down. If anything, he'd let too much freedom reign among the colonels, majors, captains, and lieutenants created by the new DOC in 1976. They had all but taken over the place.

Cabana quickly saw what he was up against, coming in from outside over a small, tight elite used to having its way with the guy in charge. Raised outside Boston, he had served his first two years in corrections as a guard at the notorious Correctional Institution on Titicut Street in Bridgewater, Massachusetts, then taken a Parchman guard job in 1972. His wife, Miriam, was a Mississippian, but he was very much an outsider at Parchman—particularly after he risked his life to stop the killing of an inmate who had shot a popular guard to death with his own gun. It didn't help that he had a vowel at the end of his name and came from a place everyone called *Massajoozitts.*

In the twelve years between stints at the farm, however, Cabana had become an almost unrecognizable transplant. He now had a strong drawl, six very southern children, and he ran prisoners in the old southern style. Cabana was as cold, singleminded, and hands-on as any Parchman warden since Marvin Wiggins, a manner he had acquired working his way up through southern prisons and state services in the 1970s. In 1976, as a probation officer in Pascagoula, he'd made an impression on future Mississippi Corrections Commissioner Morris Thigpen, which led eventually to the Parchman superintendency, a job that made him the state's top corrections official outside Jackson. But he still wasn't one of them. He wasn't country, he wasn't city, he wasn't black or particularly white, or Choctaw, or Freemason. At Parchman, the officers didn't know what he was. Neither did the inmates.

To a man, however, they banded against Cabana when he announced they were to resume farming—cotton, of all things. The inmates went "on strike" the first morning out, showing up barefoot at 6:30 for the big flatbed trucks out to the fields, claiming their shoes had been stolen. The majors, who had no love for turning the penitentiary back into a farm, staged a "blue flu," and whole sections of their guard just called in sick. Cabana, who stood five-seven and sported a two-inch bald spot and a paunch, seemed outnumbered. But he was looking at different numbers. There had been a thousand working inmates living in cages when he'd left his guard job here in 1973. Now there were five thousand, lying on dormitory beds with nothing to do but fornicate, fight, escape, and replay whatever urban nightmare had landed them at Parchman, while one of the larger farms in the state lay fallow. In 1973, some seventy-five free-world employees had run the farm; now there were over a thousand guards, and the place had gone down the tubes. Cabana also had a more personal number in mind: the fourteen years it had taken to become the boss.

He had experience with inmate labor recalcitrance as well. In 1972, there had been a slowdown by the okra pickers on the line he was guarding. Picking okra is hideous work. The plants are low and spiny and hard to pluck without getting stung. Cabana had listened patiently to those inmates' grievances—he was a great listener; then he collected their work gloves and spent the night cutting holes in the fingertips. The men had come back the following evening with red gloves, and the rows were picked clean. He admired the cleverness of these barefooted inmates, then told the few guards that had come to work to line the men in "deuces," two by two, and walk them out to the fields the long way, over several miles of blacktop. By 7:30, when the sun was up and the asphalt was toasty, 95 percent of the men remembered where their shoes were. The tougher ones spent some time barefoot out in the cotton fields, shackled to the big flatbed trucks, until their memories came back.

Dealing with the free-world officers was harder. Cabana could see the hierarchy among the top men once he had them seated across the big oak desk in his office. Gregg followed Childs. Van Landingham took orders from no one. He would have been superintendent material himself, but his indomitability made him unable to engage in the give-and-take needed to manage the 6,000 conflicting personalities on the farm. The Parkers were Parchmanites, and would go with whatever resolution was found. Hocutt, with his sad, bright eyes and a face that probably didn't look any older than in his high-school yearbook photo, was a bag of paradoxes: a loner standing shoulder to shoulder in mutiny, with rapt attention to every word Cabana uttered.

Cabana read them the riot act, then stopped in mid-sentence. These men had been the law here for years; he remembered them all from Jimmy Lee Gray's execution, when he'd watched their cold-blooded teamwork through the chamber's plate glass windows. He looked at van Landingham now, wondering how he could sway such an obstinate man. "Tomorrow morning," he finally said, making it up as he went along, "the train is leaving the station. Either get on the train or get off the tracks, 'cause the train will run over your ass. Good day, and get the fuck out of my office." Field work started in earnest the next morning.

Cabana knew he had to go outside the closed core of this goon squad to form the nucleus of his office. He picked Steve Puckett, who was on his way to the superintendency himself, and Dwight Presley and Wayne Fleming, two friends from his 1972 stint at the farm, as the deputies to oversee his administration. A tour of the MSU tiers let Cabana know that Hocutt was his man for the actual running of inmates, particularly in the fields. During his years as a guard, Cabana had put in three months at the MSU, and the change in the building astounded him. Inmates were on their bunks, reading, watching TV, listening to radios at respectful volumes, and their cells were clean. The officers were tough, cold, and bored. When Cabana had seen these tiers in 1972, the only question in his mind was whether the inmates had come in as wild beasts or been made so by the MSU. But these men looked like god-fearing Christians on football Sunday—violent spirited but alert, with a modicum of hope somehow still shining in them. There were no threats, no insults, and, Cabana realized as he and Hocutt made their way from A Tier across to C Tier, no urine, feces, or fiery objects flying at them.

Cabana had a cloistered but powerful affinity for buildings like the MSU. In a fourteen-year career, he'd worked at all manner of prisons, but like a recidivist criminal somehow always found himself working with each facility's most incorrigible inmates, and spending the bulk of his time inside maximum security buildings. It was clear Hocutt shared this attraction. He had been running between six and eight camps at Parchman for three years but kept his personal office at the MSU, despite the isolation and oppressiveness of the building, and he spent far more than his quota of time on its tiers. Only a special sort could look at such places and inmates as anything but a collective nightmare. Over the next four years, as Cabana got to know the depths of Hocutt's contempt and pessimism, he'd realize that the MSU's cells provided one verification after another for Hocutt's belief that life was hell. For Cabana, who harbored graduate-school fantasies of reform along with a mordant understanding of his job's realities, the MSU inmates were the quintessential red-headed stepchildren, incapable of reform and rehabilitation. As a warden, he had to come to terms, day in and day out, with the harsh truth that he was all they had, and the even harsher truth that he just didn't matter to them. The inmates of the MSU were an elite of their own.

Among this unique population, death row had its special, sickly grandeur. Cabana had the heebie-jeebies the moment he and Hocutt stepped through the barred double gates leading to the row. He thought it was the smell. A Tet Offensive veteran, Cabana had served as an air force paramedic, operating out of Danang, and the long-term memories that come so easily with strong odors cast spells on him. Institutional cleaning products were inseparable from the hyper-disinfection of a field hospital or medevac (a public toilet mopped twelve hours earlier could make him break out in sweat that would drench his shirt), and the ammonia solutions and fake citrus essence that the trustees had used to clean death row that morning were hell on his psyche.

But he knew something else on the row was working on him: the so-called charisma of the condemned. He'd gotten his first view of this sickly magnetism in 1970, walking into a prison social at Titicut, held along with the inmate Follies at the psychiatric hospital: a line of free-world girls, waiting for their turn to dance with Albert de Salvo, the Boston Strangler. His second taste, which came during his service as an MSU guard, was stronger and more personal. *Furman* had just come down, but the condemned were still together on what had been death row. Walking past them the first time, he was seized by an intimation that he would one day be putting one of these people to death.

When executions returned after *Gregg*, Cabana saw that the condemned prisoner's charisma grew stronger as an execution date approached. His first wardency was of a small prison in Gainesville, Florida, a job he'd gotten after finishing course work for a master's degree in criminology in 1977. Two years into the job, as John Spenkelink's date drew near and death row at the Florida State Prison at Starke began to swell (135 men by 1979), Cabana received three inmates who were en route to the row. They were the kind he hated: trashy cocks-of-the-walk who boasted of their crime—they had murdered a local businessman—because they had nothing else. In Spenkelink's last two days, they stopped abusing guards and fellow prisoners, chain-smoked nervously, ate and slept sparingly, and paced their cells as though it were their own death watch. As Spenkelink's hour approached, Cabana noticed, the entire inmate population put itself on a kind of lockdown.

In late 1983, toward the end of his deputy wardency at Jefferson

City, Missouri, Cabana came close to executing George "Tiny" Mercer, a handsome biker who had killed a woman given him as a "present" by some fellow bikers. Cabana, who had witnessed Jimmy Lee Gray's asphyxiation a few months earlier, was asked to bring Missouri's gas chamber up to speed. It was unlikely Tiny would be asphyxiated. Missouri's chamber leaked badly when they tested it with a smoke bomb, and Mercer still had good chances for a stay. If it were to happen, however, Cabana was assigned to mix the sulfuric acid bath, then shoulder "one-third" of the coup de grâce: Missouri had no executioner, and Warden Bill Armontrout had decided his two deputy wardens should join hands with him on the lever. When Mercer's 1983 date was stayed, it came as more of a relief than Cabana had expected. He liked Tiny. Everyone did.

And he hated gas chambers. The image of Susan Hayward in *I Want to Live,* strangling blindfolded in the wide-open constriction of San Quentin's chamber, had given him nightmares as a kid. The first time he'd seen Parchman's chamber, on his last day as an MSU guard, he'd walked straight through the oval doorway, "moved—as if by some unseen force" and went up to touch the black chair, as though in an out-of-body experience. His day's assignment was the chamber's upkeep, and despite his dread, Cabana spent the first hour making jokes with a fellow guard, taking turns posing in the chair, and throwing the lever. Then he'd begun to feel grim and exhausted, and he was a psychic mess when the day was over. It was the chair itself that unnerved him. Its blackness, height, straps, and rigidity captured his imagination. Ten years later, as he took his seat in the witness room next to Eddie Lucas before Jimmy Lee Gray's asphyxiation, he'd felt seasick when he saw the chair, so black and inert in the brightly lit chamber.

Cabana happened to know Jimmy Lee Gray, and he despised him. They had met in 1976 in Pascagoula where Cabana worked as a corrections counselor. At the time, he had three-year-old twins, the same age as Darèssa Jean Scales when she had been raped and smothered. Gray, who was hypersensitive to anyone's feelings about him, saw Cabana's contempt before a word had passed between them. With his impassive ways, Cabana was like a Rorschach test: People had a way of saying what was on their minds when they looked at him. "You know," Gray said, "I have a strong feeling you really want to pull the switch on me yourself."

Cabana couldn't imagine a parent who would not volunteer to kill Gray with his bare hands. "Well hey, Jimmy," he said, patting him on the shoulder with gentle irony. "Maybe it'll all work out." He'd heard tales about the suffering that went on in that chamber, and the abstract prospect of watching Gray strangle struck a sympathetic chord. It was a different thing, however, to watch it happen.

As Cabana and Hocutt neared the end of C Tier and the death house door loomed ahead, Cabana suddenly turned on his heels, leaving Hocutt behind. He was halfway back to the double gates at the head of the tier when a voice said, "Warden." It came from a thin, handsome young man with a pencil mustache, a ten-gallon Afro, and a personable, peaceful smile. "Would you consider changing a rule?" he asked. "So the row could receive Christmas packages?"

"Damn." Cabana laughed. "You don't waste much time, do you?"

"Just trying to break you in early."

The inmate's name was Connie Ray Evans, condemned in 1981 for murdering a convenience store clerk during a robbery. Despite his proud, relaxed demeanor, Evans had come within forty-eight hours of the chamber eight months earlier. He was the type of inmate Cabana liked: individual but respectful, mindful of his place as a prisoner but intelligent enough to find a way to reach past it without offending his jailer or debasing himself. Cabana, who kept his own respectful distances, decided to stand on ceremony and told Evans to put the request in writing, then continued down the row.

Evans had made his impact, though, and when Cabana toured the MSU thereafter, he stopped at his cell again and again. With the death house door fifteen feet farther down the tier, Cabana gladly lingered for some death row shop talk—life and death, punishment and redemption. The row had few atheists, but Connie Ray Evans's religion wasn't just jailhouse. He was truly searching for something: in Christ, in himself, and, finally, from Cabana, whose visits he relished. After a few months, Cabana realized he was looking forward to them as well, and that Connie Ray Evans had become a friend, however inappropriate the word seemed. On a summer night three years later, they would get to know each other far more intimately.

Cabana knew it took more than one man's capacity as a jailer to make these kinds of changes in the MSU, but his instincts about Hocutt were strong, as was his desire to groom him. Hocutt was an extremely violent man, but his violence was a tool he used rather than gave way to. That kind of control was rare, and it was the *only* answer to the Catch-22 that came with running a southern prison farm. You needed a pit bull to run interference, someone with a bottomless capacity for brutality, but in using that force you almost invariably became what you were trying to control. Hocutt had the capacity to hold it in just short of lawlessness. Equally important, his ears were open: to things like orders, to men eager to obey them, to the obscure argot of the convict—everything from a man on the verge of losing it to nascent corrigibility—and he had a good administrative nose for other men with the same cold-blooded intelligence.

Hocutt and Cabana came from radically different places, but they had deep similarities in their approach to work and people. They identified with their jobs far too much, and were equally unable to leave them behind when they went home. Both men gloried in nondescription, had a passive-aggressive genius for giving people all the rope they needed, had dry wits and blank smiles, and saved their deepest contempt for affectation and cud chewing. Hocutt and Cabana spoke simply and to the point, though they did share the unique tic of lapsing into bad, jokey colloquialism when addressing aspects of the job that were less than admirable—generally, anything having to do with institutional violence. That part of their jobs, and a corresponding inability to relax, had literally shaped both men over the years. In profile, they looked like question marks—rounded backs, hunched shoulders, heads bent forward over their bodies to an uncomfortable-looking degree. In conversa-

tion, both men swayed slightly from side to side, or rocked gently
if they were sitting, their arms crossed tightly on their chests and
their hands balled into fists under their armpits.

Hocutt, who despised the bureaucratic mess Parchman had
become, saw Cabana as the last of the true southern wardens.
Cabana needed a man who went back far enough—in spirit as well
as years—to run Parchman as a farm, but who was young enough
to be free of cronyism. He told Hocutt that he also needed a band
apart—an elite corps outside the majors and colonels. Hocutt put
together a fearsome bunch of country boys: thirty-five officers,
eight sergeants, three lieutenants, and himself as field major. There
was no money for the uniforms he would have loved to dress them
in, so everyone just wore the same clothes, day after day—blue
jeans, blue shirts, dark boots, and white hats—and became known
as the White Hats. They all rode horses, except Hocutt, who requisitioned a white GMC pickup. Most of the officers drove black
trucks, but Hocutt liked his white so the filth, dust, and muddy soil
(known in the delta as gumbo) that he spent his twelve-hour days
speeding and splashing through would show to advantage. Cabana
loved the look, got a white hat and a white GMC himself, and spent
the bulk of his superintendency covered in mud.

For the next four years, the White Hats shook inmates from bed
and ran them to the fields and back to the camps for the nine months
each year that the prison farmed. Outside of a twenty-minute lunch,
hailstones, or hurricanes, the work never stopped. If an inmate had to
urinate, he did it on line. If he had to defecate, he took a six-foot pole
with a flag on it to a patch that had been worked and kept the flag
waving, so the White Hats would know he hadn't bolted. In winter,
they shook down units randomly for contraband, and used extreme
shows of force to quiet disturbances that arose whenever the men
weren't worn out from twelve-hour days bent over a row of cotton.
Generally, all it took was the sound of hoofbeats, or the sight of them
coming over a ridge or field. "Shit," you'd hear. "Here come those
fucking cowboys." The White Hats ate together, drank together,
and took vacations together at hunting season, losing two weeks to a

roaring bender in the woods of Scoby County, euphemistically called Deer Camp. After fifty straight weeks running inmates, no one had any interest in killing some defenseless deer.

In April 1984, six months before Cabana's superintendency had begun, the Mississippi State Legislature had ratified the first changes in the death penalty protocol since 1954, spurred by the controversy Gray's head-banging had aroused. Anyone condemned after July 1, 1984, was to be executed by lethal injection; those condemned prior to that date were "grandfathered" into the chamber.

This move to the gurney had come relatively early (Mississippi was the eighth state to adopt lethal injection) but not easily. Senator Con Maloney, the author of the lethal injection bill, had been pushing it since Oklahoma had made the switch seven years earlier—largely, he said, to save on the chamber's upkeep, which he estimated, with a bit of English, at $20,000 annually. Maloney had reintroduced variations of his bill in every ensuing year, but the legislature was deeply wedded to their chamber, which for thirty years had been a significant icon down in Jackson. There were spin-offs and subplots to the debate, but behind them all was the question of Mississippi's contemporary reason for executing, and that issue divided hard-liners and reformers more than anything on the legislative calendar: Did you execute to send a message or to dispose of a problem? The question found form in the method. Did you put a man to sleep or did you kill him? Representative Jerry Wilburn of Mantachie felt the gas chamber was *too* mild, and suggested a return to public hangings. "It sounds a little cruel," he admitted, "but it's time we let criminals know we mean business. What's wrong with hanging a thug?"

Senator Howard Dyer of Greenville, a former prosecutor and the legislature's leading champion of the chamber, had used graphic details to paint Daressa Jean Scales's suffocation at the hands of Jimmy Lee Gray when he'd argued for retaining the chamber at the beginning of 1984's debates. A group of third graders visiting the senate gallery were glued to the balcony rail for his summation-style peroration, and the *Clarion-Ledger* story the following day about the horror on their tabula rasa faces had not helped Dyer's case. If suffocation was torture, as Dyer seemed to be saying, was Mississippi no better than Jimmy Lee Gray?

"I cannot conceive of any method of taking a life that is not unpleasant," said the Senate Judiciary Committee's chairman, Martin Smith, of Poplarville, before voting against lethal injection in the first round of ballots on Maloney's bill, in February 1984. Dyer was more direct in casting his nay when ballots were taken again a month later. "Let's leave well enough alone." The courts, he argued, had no problems with the chamber.

He was wrong. The Fifth Circuit, which had been so explicit in affirming that the chamber met the Eighth Amendment test of cruel and unusual punishment in re *Jimmy Lee Gray*, 710 F.2d 1048, were clearly harrowed by the graphic accounts of Gray's asphyxiation, which had made him a cause célèbre for abolitionists worldwide. That judicial embarrassment became evident when the other Fifth Circuit states and the rest of the Death Belt began to execute almost blithely, while Mississippi failed again and again to send its condemned to the gas chamber. Louisiana, after some initial judicial tentativeness (two men had reached death watch before the warden was ordered by state courts to stand down and return the prisoner to death row), electrocuted four men in 1984; Florida killed eight; and Texas was off on its phenomenal pace, executing roughly ten men per year. Like Mississippi, Alabama, the only other Death Belt laggard in 1984, had suffered a botched execution in 1983, that of John Louis Evans III, four months before Gray. The courts were making it clear: We cannot sanction an execution under stigma or perception of cruel and unusual punishment. Eight years had now passed since *Gregg*, Mississippi had brought ten men within the shadow of the chamber, but the state had only Jimmy Lee Gray's execution to show for it.

One particular last-minute stay by the Fifth Circuit, at the end of February 1984, proved a watershed for Maloney's lethal-injection bill. Marion Albert Pruett, the one-eyed "Mad Dog," was a North Carolina–born serial killer who had shot, stabbed, and raped a dozen people, killing six, on a drug-induced rampage across I-10 from Alabama to New Mexico. Three states had already condemned him when he landed on Parchman's death row in 1981. His longevity there, a running embarrassment for the state, was felt most keenly after the February 1984 stay, as Pruett's Mississippi death sentence had been affirmed only weeks before in a sharply worded Supreme Court denial of his appeal. When the Fifth Circuit overrode the higher court and stayed Pruett, the finger was pointing directly at Parchman's death house.

To compound matters, Pruett had begun to glory in making himself a poster boy for the death penalty. "Mad Dog" was a tag he'd unintentionally stuck on himself by way of exculpation ("Drugs made me a mad dog," he'd told a reporter in 1981), but the press had made a moniker of it, and Pruett had quickly learned to enjoy the notoriety it brought. Each new interview presented a fresh chance to express the joy he derived from his victims' horror and suffering, of using his missing eye to mug for the cameras, of assuring the media he would not hesitate to kill again. It was silly, insidious theater, a grotesque of media profiteering and collusion in death penalty controversy, but it humiliated the DOC, the attorney general, and the pro-penalty men of the legislature. The point got razor sharp in March, when the attorney general in Arkansas, where Pruett was also condemned, pressed Mississippi for extradition, to show that his state had no qualms about taking care of business.

At the legislative session following Pruett's February stay, the hardliners gave way like a dam bursting. After failing to reach "acceptable form" for seven years, Con Maloney's bill required only three hours of rewrites in committee to get the full senate seated. The final tally was 34–8, the yea votes bringing some truly unctuous plainspeak. "Dead is dead," said Senator George Guerieri. "I'm for it."

"[H]ang 'em, gas 'em, or inject 'em," Dyer relented. "It makes no difference to me, just as long as we remove 'em."

Some items got lost or softened in the pork barrel. A raise of the executioner's fee to $1,000, mandated in Maloney's original bill, was limited to $500, though the executioner was now officially allowed a second deputy in the chemical room, at a rate of $250 per execution. Allowing the prisoner to choose between the old and new methods, a measure adopted by most states in their moves from chair or chamber to the needle, was erased. A rote borrowing from other states that had made the move, it aroused strong objections in Jackson, where legislators were loath to give up the right to asphyxiate men already condemned to the chamber. Instead, they set the random date of conviction of July 1, 1984, after which the condemned would die by the needle.

A room behind the gas chamber's witness room was cordoned off to form a lethal-injection suite, and Hocutt and the other members of the execution team were trained to establish catheters and inject the three drugs in stages. The suite had its own chairs for witnesses—ascending rows of six seats apiece, which to Cabana looked like a jury box. Ten years and two executions later, when he had left corrections work and become an abolitionist, he would propose that juries sit as the witnesses of all executions, with eleven serving as witnesses and the foreman joining the district attorney and judge in pulling the switch—perhaps a memory of Missouri's proposed plan to have the warden and his two assistants pull the lever of the gas chamber. To Hocutt, who felt executions should be widely observed, if not televised, the ascending rows just looked like a movie theater.

In the lethal-injection bill's final centralizing measure, the county sheriff would no longer "run," or even attend, executions. The county of conviction would now be represented only by a limited press corps. Instead, all executions would be formally overseen from Jackson, with Corrections Commissioner Morris Thigpen having final say in all arrangements. One of his first actions was to grant a BBC documentarist's request to film the state's next execution protocol, should the condemned give consent. That now made for a thorny issue. No one doubted that the Mad Dog would be the first to go and, given his camera craziness, that he would give his consent. In fact, Pruett was already intimating through his lawyers that he had the right, under the First Amendment, to have his execution videotaped. How many stays would that mind game attract?

On a trip to the MSU, Cabana stopped by Pruett's cell to look at the infamous man and to play some mind games of his own. "Hey," he said. "Tell me one thing. Why didn't you kill children? They're fair game, aren't they? Tots. Babies. Newborns?"

Hocutt, by Cabana's side, knew the warden was taunting Pruett, but he also knew he'd taken the wrong tack.

"Pruett," Cabana tried again, "we've got a population explosion in this country. The last woman you killed was in her twenties. Why didn't you go for the little ones? I mean, what's your problem?"

"I was six months old the first time my father bounced me off the floor and fractured my skull," Pruett said. "I kinda got a thing for little kids, Warden. When they hit sixteen, though, bring 'em on by."

Cabana had heard more piteous child abuse stories, but he had to admit the delivery was excellent. Marion Pruett, he saw, was crazy like a fox. Like Florida's Ted Bundy, who had inspired a fashion for electric-chair lapel pins, frying-pan rearview-mirror ornaments, and such bumper stickers as TED: THANK GOD IT'S FRYDAY, and I'LL BUCKLE UP WHEN BUNDY DOES (in reaction to Florida's controversial mandatory seat belt law), Pruett was shrewdly using the excess of his crimes to stay alive. The Mad Dog, Cabana knew, would not be the man he first executed.

Hocutt saw Cabana hesitate and mistakenly assumed that Pruett had won the cheapest form of death row emotional blackmail. Everyone in here had stories like that, and for Pruett to have snuck that head-bouncing shtick past the new boss made Hocutt wonder if Cabana would hold up when the time came. The new warden had ice in his veins, but killing with premeditation took a different kind of coldness. At the end of the day, you had to *want* to kill. Hocutt had wanted to kill Jimmy Lee Gray.

He also wanted to kill Pruett. The man's posturing disgusted him, though he kept it to himself. Death row inmates didn't mess with their jailers, and Hocutt had an inviolable quid pro quo about that. But the Mad Dog routine made him want to punch the man's face with both fists, maybe even hit him with a stick a few times. Instead, every month or two he had to walk Pruett to the MSU's back fence, so another TV crew could film his song and dance.

The next crew that came after the meeting with Cabana was from Albuquerque, where Pruett had also killed. The cameraman and interviewer soaked it up, and Pruett was so full of swagger Hocutt couldn't hold back any longer. On the way back to death row, he decided he'd taunt him about how bad his old man's aim was when he'd bounced his head off the floor. When he closed the door on Pruett's cell, however, he opted instead for a little MSU soft soap.

"You are cold, aren't you?"

"You know it, too," said Pruett.

"*Damn* good thing we got you behind bars. *Damn* good thing."

"Long as you can keep me, fat boy," Pruett smiled. "They want me bad in Arkansas."

Hocutt shook his head. "One day soon, you're going to be looking out the window of that chamber at the end of the hall. That's when you'll know what cold is."

"What, you?"

"Take a look."

Pruett shot the Mad Dog sneer. Hocutt took the cue and came a step closer as Pruett gripped the bars. When they were as near as he cared to come, Hocutt made an O of his lips. "Woof, woof," he said.

Commissioner Morris Thigpen left the Mississippi DOC for a similar position in Alabama in December 1986, and Cabana was given the top job on an interim basis. He named his friend Wayne Fleming the acting superintendent at Parchman for the two or three days each week he would be in Jackson, and he made Hocutt a colonel. With Hocutt's promotion came one of the big houses near the prison's front gate, a few doors down from Cabana's. Also allowed his own personal trustee now, he chose a huge, semi-retarded man from the southern piney woods named Katerhoochie. He was serving life for murder, but Hocutt was amused by him, and knew he was just a gentle giant who had gone haywire after brewing a pot of tea from psilocybin mushrooms. He had destroyed the chandelier in his grandmother's house, thinking it was a glass butterfly sent to kill him, and when he was finally arrested, driving 75 mph down I-10, the remains of a man were chained to his bumper. The strange choice of sidekick increased Hocutt's wild-man image in the farm. In addition to his mud-caked truck, boots, and white hat and the unofficial executioner's job, he now had a certified lunatic by his side.

Cabana's job as interim commissioner kept him shuttling up and down 49W between Parchman and Jackson. Halfway down, at the four-way stop at Highway 12 (the famous crossroads where Robert Johnson "sold his soul to the devil" for his guitar genius), Thomas Berry Bruce sold vegetables at his roadside stand. Cabana often stopped for fruit or squash and chitchat. Bruce seemed drunk much of the time, and quiet as always when he drank. Cabana wanted to do the talking, anyway; he was on a mission. He wanted Bruce out of the picture. He had seen him in 1983 through the chamber's windows, throwing the switch on Jimmy Lee Gray, and Bruce had looked stoned to him. The execution team swore up and down he'd been sober, but Cabana guessed they simply didn't want to break ranks. Barry Parker, who had seen Bruce at the guest house the afternoon of Jimmy Lee Gray's asphyxiation, said the old man was so drunk he could barely stand.

True to form, Cabana just asked questions: about the old days, about the persimmons, about whether Bruce had ever thought about giving up the executioner's duties, maybe taking on a younger man, a southern-style apprentice. At the crossroads, however, Cabana only learned that subtlety was the wrong tack with Bruce. He couldn't even tell if his nuance went above Bruce's head or below. All he got in return was a simple, redundant message: Thomas Berry Bruce ran Mississippi's gas chamber, and had since 1954.

In Jackson, Cabana made friends with Marvin "Sonny" White.

A short, balding man with a paunch, White was the assistant attorney general in charge of all death penalty matters. Both men hated the media and looked upon executions with fear and revulsion—an unusual stance among top Jackson bureaucrats, who usually fought for that kind of prominence in the public eye.

The Supreme Court had reconvened, and their list of impending cases gave Cabana and White much to discuss. The Court had agreed to hear the death penalty appeal in a famous case, *McCleskey* v. *Kemp,* and their decision was going to affect the Death Belt enormously. *McCleskey* had become a magnet for abolitionist professors and lawyers who helped document the appeal of cop killer Warren McCleskey. The crux of the appeal was what had become known as "proportionality": One of *Furman*'s principal grounds, back in 1972, was the disproportionate numbers of blacks who had been executed. Death Belt states had taken pains to address this. Following *Gregg,* sixteen of the first twenty executed were white. By 1987, however, the numbers seemed to show a new sign of racism: the race of the *victim* of the capital crime, rather than the man convicted for that crime, had become the new color line. In the ten years since Gary Gilmore's execution, there had been sixty-seven men and one woman executed, but in only five cases had the victim been black, a number completely out of line with the high percentage of black murder victims. If *Furman* had made clear that, by constitutional decree, black life could be no less precious than white, how then to explain this disproportionality in post-*Gregg* figures?

In March, the court came to the rather shocking decision that they just didn't have to. *McCleskey*'s claims of disproportionality, they ruled, however well substantiated, showed only a "discrepancy that appears to correlate to race." A coda provided their clearest pro-penalty mandate since *Gregg:* "[S]uch apparent discrepancies are an inevitable part of our criminal justice system." One could hear the death house doors from Florida to Texas yawning open, particularly at Parchman: of the thirty-plus men then on the row, not one had been convicted for murdering a black man.

Sonny White spelled it out for Cabana. "Within a year," he said, "you're going to be running an execution." The most likely to go, he told Cabana, was his friend Connie Ray Evans, for whom the timing of *McCleskey* couldn't have been worse. Evans had all but run through his appeals since the 1984 eleventh-hour stay that had forced the state to stand down from his asphyxiation. His murder victim, an Indian, also made him an ideal choice. It was the closest Parchman's death row had to offer to a "colored" victim.

But it was not to be Connie Ray Evans. On March 30, just weeks after *McCleskey,* the Mississippi Supreme Court denied the appeal of the young man in the cell next to Evans. His name was Edward Earl Johnson, and the timing of his appeals was even worse than Connie Ray's, as his crime and particulars bore an unfortunate resemblance to McCleskey's. Both condemned men were black, and both had been sentenced for murdering a white policeman in the late 1970s while committing a felony robbery. On a constitutional level, their similarities went even further: after *Gregg,* prosecutors had to prove to juries in capital trials that the crime specifically merited the death penalty because of some "aggravating circumstance." That nebulous quality, which varied widely from state to state, had been an important issue in *Gregg*, as it was meant to address the penalty's arbitrariness by codifying exact grounds for a death sentence. In practice, however, its application had proven rather ad hominem: various district attorneys pursued certain aggravating circumstances while ignoring others, and jury discretion on the matter seemed subject to local prejudice. In both Mississippi and Georgia, for example, attorneys general knew that murders of policemen in the commission of a felony robbery were particularly "aggravating"—to juries and newspaper readers—and they pursued a death sentence in almost every such case.

Johnson's lawyers had unsuccessfully argued, from the Fifth Circuit up to the Supreme Court, that the doctrine of aggravating circumstance was arbitrary and therefore in violation of the Eighth Amendment. All that was left them, after these denials, was a final hearing on that issue with the Mississippi Supreme Court. When that court denied Johnson in March, and issued an execution date of May 20, there was little doubt left about the young man's fate.

At their meeting on April 1, Sonny White told Cabana he'd been off by a few months when he'd given it a year to happen. Then he paused and told Cabana to brace himself. "I'm getting word there'll be another one after Edward Earl," he said.

"Who?" Cabana asked.

"I'm guessing Connie Ray Evans."

"Why Connie Ray?" asked Cabana. There had been no issue of aggravating circumstance in his defense.

"His time's up," White said simply, then shrugged his shoulders.

There was a long-standing Parchman covenant for the governor to spend one night each year at the warden's guest house, after which he would inspect the prison. This year's visit, from Governor Allain, came a week before Edward Earl Johnson's final appeal failed.

Allain and Cabana had always spoken warmly over the thirty months of Cabana's superintendency at the farm; since December, when Cabana had taken over in Jackson as well, the two had grown much closer. Mississippi Catholics are few and far between, and both men were at pretty much the same spot in the middle of the road as far as religion was concerned—they were the types who never missed Sunday mass but also never quoted scripture or even voted along religious lines. They shared a professional friendship with Bishop Houck of the Southern Bishops Conference and had spoken with him frequently about the church's increasingly dogmatic abolitionist stance under Pope John Paul II. Houck was also personally against the penalty, but he understood the difference between doctrine and the rank and file, both of his priesthood and his congregation. It was a rare Sunday that passed without a *Roe* v. *Wade* reference, but you never heard the names *Furman*, *Gregg*, or *McCleskey* from the pulpit.

Heading down Parchman's guard row in the governor's limo for the prison inspection, Allain turned to Cabana in the backseat and said he had a kind of clemency to offer, though not for Johnson. "The state supreme court hasn't come down," he said, "but Sonny tells me they've got this kid." Allain said he'd given it much thought and wanted to assign Cabana to a big project in Jackson, starting in early May. After all, he was corrections commissioner, and that was a Jackson position. There was no reason for him to run an asphyxiation in the delta. "I want you to let a deputy warden do it," he said.

"I'll be happy to," Cabana said with typical irony, "if you'll deputize your lieutenant governor to make the clemency call on Edward Earl." He knew that Allain could never take such a step.

The governor dropped the issue, but Cabana had one of his own to raise. He'd tried once by phone, gotten nowhere, and this seemed the time to try again. "Bill," he said, "I want to ask you to consider clemency for Edward Earl."

At issue, both men knew, was the man's guilt. Cabana had been through Johnson's file countless times, and was convinced equally that he *was* guilty, that he had been given due process, and that his own feelings in the matter were irrelevant. Straight down the line, however, the row was convinced Johnson was innocent, and that he had been framed and forced to confess by the police of Leake County, where he had been convicted of murdering a town marshal in the line of duty. Johnson had maintained those claims without variation throughout his trial and incarceration, as well as in his infrequent conversations with Cabana.

In the back of the limo, the two men talked it over. The convict was a strange one; so was the crime. Well after midnight on June 2, 1979, a seventy-one-year-old white woman was threatened with rape at gunpoint by a man who may or may not have been Johnson, who may or may not have come at that strange hour to pay for some Avon products that had been ordered by a relative. Just before sunrise, a marshal was found dead on a nearby lawn, shot five times in the head. Three of the bullets came from a .25-caliber pistol, two were from his own .357 Magnum revolver.

Johnson's car had broken down in the area late that evening, and phone records of his call for a tow truck led police to question him. He was interrogated and brought to the house of the elderly woman who had been assaulted, accompanied by his grandmother, who had raised Johnson from an early age and still kept a close eye on him. He also stood in a line-up for the assault victim's tenant, who had witnessed the crime. The elderly woman said it hadn't been Johnson who'd made the threats, the tenant failed to pick him out of a line-up, and the police let him go.

But he was arrested again two days later, and this time his grandmother was not allowed to come along. Brought in for a second line-up, the assault victim's tenant was able to positively identify him—not surprising, inasmuch as he'd stood in a totally different line-up two days previous—and this time, the elderly woman said that Johnson *had* been the man who'd attacked her. En route to Jackson, or so he later claimed, Johnson had been marched out to a woods and threatened with his own death and the death of his grandmother unless he confessed to shooting the marshal. He claimed he had also been made to "lead" police to the two pistols and forced to sign a full confession. Down in Jackson, he retracted his confession the moment he was released by the Leake County sheriff to Jackson officials.

The oddness of the crime and its investigation haunted Cabana. After seven years on two death rows, Cabana had heard odder stories, as indeed had Allain in his four years as attorney general. It was the Parchman row's unanimous vote on Johnson's innocence, however, that led Cabana to ask for clemency. He had never heard anything like solidarity from there before. Those men were like atoms or monads, sitting in individual cells year in and year out, and they didn't cast their ballots lightly: each man on death row had one plea of innocence in mind. Johnson was hardly their type either. There were other quiet people on death row, but he was particularly enigmatic: short, with thick glasses, well groomed, with a measured, delicate voice (his nickname at home was Squeaky); prior to his capital conviction, he'd had a completely clean record, good employment history at the local poultry plant, and not an ounce of street to him, which was more than rare on the row. Several of his legal challenges cited early brain damage and a low IQ, but he clearly had mother wit and the sharp, interior logic of the loner. Edward Earl Johnson spoke in well-formed sentences, read all day long, and played a decent game of chess.

It was that indistinct, idiosyncratic manner that made it most difficult to judge his culpability. As a former prosecutor, Allain wondered if Johnson's jury had had the same problem. Most of the defense lawyers who had handled Johnson's case were said to be bewildered, and they included a future state supreme court justice Allain had appointed. To them, it really was a coin flip. To Cabana, Johnson's presentation was also somewhat unclear. At times, he seemed to be saying some informant had been able to plea down to some lesser charge for fingering him; at others, that some larger conspiracy was at play, that the sheriffs had, as he said, "simply taken the first nigger they could find." The only constant was that his execution was a legal lynching.

This "southern justice" alibi had little effect on Cabana. "C'mon," he'd say each time. "This is not the nineteen fifties. Black folks know how to get to the Justice Department, the FBI."

"You don't know what these people would do to my family," Johnson would say, and go quiet. Cabana would stand outside the cell, wondering where to take the conversation next. He'd realized then, and he saw once again as the limo headed down guard row, that the young man's enigmatic silence was going to be a great help to him in getting through the execution. It reminded him that guilt and innocence was not the concern of his conscience as a warden. That would be served by not taking the easy way out, by being there for his prisoner when his time came.

Allain, on the other hand, was a politician, and politicians sold the public on their conscience. He also happened to be a man of private conscience, too much to hide the realities of his decision from Cabana. "It would be impossible to do what you've asked," he finally said. He had come to office on an execution's aftershock and as state attorney general had handled many of Mississippi's responses to Johnson's appeals before the Fifth Circuit and the Supreme Court. "Connie Ray Evans's as well," he added, "should that one come to pass." Neither man mentioned it, but there was another reality at work too. The state's constitution, a Reconstruction-era document drafted by a legislature determined to limit the power of the governor, had been amended recently to allow the incumbent to run for a second term. Allain wanted to run again, and it was less than a year from the primaries. Johnson was a convicted cop killer.

Cabana had another pressing subject. Mississippi's executioner, he reminded Allain, serves at the pleasure of the governor. "Berry Bruce," he said, "isn't taking any hints."

"Get him to quit," Allain said wearily, "or I'll fire him." Cabana could see the conversation had worn Allain out, and they still had a long day ahead, inspecting the farm. Before they got out of the limo, Allain turned to Cabana one last time. "Don, please don't make me fire the old guy. You can't imagine how hard that would be."

On Cabana's last trip to Jackson that April, en route to a corrections meeting in Biloxi where he would see through details of Johnson's execution protocol, he pulled over on the corner of Highway 49W and 12 one last time. He put the question bluntly to Bruce: "Will you retire?" Getting no response, Cabana conveyed what the governor had said. Bruce's face colored, but he still said nothing.

The following Thursday, five days before Edward Earl Johnson's date, the Jackson attorney general's office announced Bruce's resignation. "I'm just tired of the whole thing," Bruce was quoted in the *Clarion-Ledger* as having told the governor in a private phone conversation. "[L]et some younger fella fool with it."

CHAPTER SEVEN

"I DON'T WANT TO DO THIS ANYMORE": THE EXECUTIONS OF EDWARD EARL JOHNSON AND CONNIE RAY EVANS

DONALD HOCUTT, CABANA DECIDED, was the officer best suited to execute Edward Earl Johnson. He was of sufficient rank and station now, regardless of whether van Landingham, Childs, Gregg, and the Parkers would feel stepped over. Then he reversed himself, deciding he didn't want any prison personnel touching that lever, and tapped the Mississippi Highway Patrol investigator who worked at, but not for, the prison—a strong, silent type named Charles Tate Rogers. Governor Allain made the appointment the next day.

After thirty months, the colonels and majors knew Cabana had ideas that couldn't be shaken, and that this was one of them: killing inmates was a violation of their jobs as correctional officers. The idea was absurd to every man on the execution team, and they spoke on it in turn, but Cabana's only concession was to let van Landingham be the one who put the cyanide below the chair.

The appointment of Charles Tate Rogers didn't come up in Cabana's conversation with Hocutt. Instead, Hocutt raised the issue of Gray's head-banging. They weighed their options. The metal pipe couldn't be taken out; moving the chair, even a half foot, would mean overhauling the entire plumbing system. That left redesigning the chair, tying the man's head to the pole with a belt, or dressing him in a football helmet.

Or they could simply pad the pole. Hocutt went to the hospital to find some institutional device, came back with the headrest of an old dentist's chair, then had a leather shop trustee attach straps and a leather cap to it. Cabana said there would probably be others after Johnson, so Hocutt bolted it on the pole with an adjustable fitting that let it slide up and down according to the man's height. It was tested on an officer named Shep Haga, who was Johnson's size; Haga couldn't move an inch. The team also used the opportunity to practice taping the wires to the stethoscope and an EKG on Haga's chest, and when Cabana stepped into the chamber he decided he might as well practice reading the death warrant. "Okay," he said, trying not to look at the chair as he mimed unfolding paper. "Say your good-byes, Shep!" Then he stepped out, the chamber was shut, and they ran through a dry practice, not even using placebo chemicals. Cabana watched it with simulated dispassion, smiling at his own graveyard humor, until he saw the EKG. Haga's heart rate was high and erratic, making crazy green wave patterns on the black screen. From that moment, Cabana became terrified about Johnson's asphyxiation. At meetings of the execution team that he called daily, he'd rock with his arms crossed, asking question after question about the protocol and the chamber's mechanics.

Requests to witness the execution started coming in; he ignored them at first, then, remembering his own experience as Jimmy Lee Gray's witness, ordered black drapes put on the three windows in the witness room, which would stay closed on the night of execution until he ordered them opened. As the final week approached and the media requests began pouring in, he decided on a lottery for the few seats available, and dozens of names went in the hat. The media were up in arms: word had leaked out that the BBC documentary team Thigpen had agreed in 1984 to give access to had obtained Johnson's consent; they had been filming since May 3 on locations denied other media. When it was learned that the English film crew had also been staying inside the prison walls, at the guest house no less, there was a sickening clamor, particularly from one national TV personality who forced his way into Cabana's office, demanding his own interview with the condemned.

"You're absolutely right, and I apologize," said Cabana. "I'll ask the condemned right away if he wants to speak to you." Of course, Johnson had no desire to.

Local corrections people and Parchman staff wanting in were harder to handle. Charles Tate Rogers came with a request from the sheriff of nearby Drew, who wanted to attend the execution. Cabana felt the Drew sheriff had no business being there but reluctantly agreed. After all, the man asking the favor was pulling the switch. He put up more of a fight in an argument over his decision not to use Fred Childs on the execution team. Hocutt and Barry Parker put up a relentless case for their partner, however, and Cabana finally caved on that too.

The BBC documentary team was filming ten hours a day—in the fields, in Cabana's white GMC, in protocol meetings, the chamber, the chemical room, by Johnson's cell, in the laundry-kitchen behind the death house. They were courteous and deferential to the realities of the row, both to inmates and officers; to Hocutt they sounded like they were talking underwater. On the morning he was headed to the K-Nine unit for the rabbits that had been rounded up for the live tests, he was perplexed by one assistant director's determination to learn about the MSU's inventory. The crew were filming behind the death house when Hocutt passed by. He'd been measuring out 500-milliliter bottles of distilled water in the chemical room.

"You have rather a lot of stuff here, haven't you?" asked the assistant director, pointing to a shelf of bleach and tomato puree.

"What kind of stuff do you want to know about?" Hocutt asked.

"Your . . . stuff," said the man. "You're well stuffed here on the row. Why do *you* suppose that is?"

"Got me," Hocutt said. "Listen, I gotta go see a man about some rabbits." It was only later, when he'd learned the accent, that he realized the assistant director had been pointing not to the supply shelf but to the tiers beyond, asking about MSU personnel, or *staff*.

As per the protocol, the execution team had to perform two live tests using animals, and Hocutt had ordered that four rabbits be rounded up. At the K-Nine unit, however, he was told by the sergeant that they hadn't found four rabbits, only two big black ones. "I got a turtle," said the sergeant.

The first test was an hour before dusk. The windows between the death house and C Tier were slammed shut, and the death house was cleared of everyone except the men performing the test, the BBC film crew, Cabana, and Deputy Warden Dwight Presley. Hocutt, who had chosen Barry Parker to be the second deputy in the chemical room, mixed the sulfuric acid. Again, they wore rubber aprons, boots, gloves, and gas masks. It was oppressively familiar: the stopwatches, timetables, the twenty-seven-step protocol, the formality, the jokes, the brownies. The two black bunnies were put on the big black seat, and the cyanide crystals were dropped in the well by van Landingham. Fred Childs, standing in for Charles Tate Rogers, shut the door and dropped the lever. The rabbits fought and frothed a few seconds, turned over, then their legs went up. Everyone shrugged their shoulders.

Except for Cabana, who was shaken by the ammonia stench when the chamber door was opened. Chain-smoking with Presley in the darkened halls, he couldn't hide his dread. "Jesus Christ," Presley said, watching the sweat roll off Cabana. "Can you imagine what the hell that thing does to a man?"

The next evening they learned what it did to a turtle. Again, the windows were barred, the rubber went on, the sulfuric acid was mixed, the cyanide went in the bowl, and van Landingham dropped the lever. The turtle retracted its head. Cabana and Hocutt gathered on opposite sides of the chamber with arms folded, watching the inert shell in the big black chair, unsure if it would flip over like the rabbits or if its legs would go up. Five minutes passed, then ten.

"I think he can't flip because his shell's so heavy," suggested Gregg, who stood next to Hocutt.

"He is built like an army helmet, ain't he?" said Hocutt.

They evacuated the chamber, Van Landingham went in, lifted the turtle off the chair, and dusted it down. All six men went in and examined the carapace from top to bottom. After a few minutes, the leathery little head came out, and the turtle made a quick flick of its red tongue.

"That's a tough little sumbitch," van Landingham yelled.

"Maybe he held his breath?" asked Gregg.

"You cannot gas a turtle," said Hocutt. "Now we know."

They carried it to a nearby stream and watched it swim off. "Tough sumbitch," said Hocutt.

The turtle test made for Cabana's first and only problem with Edward Earl Johnson. The smells wafting down C Tier brought emotion from him no one had seen before. He just couldn't believe they would use animals to test a machine being geared up to kill him. "I hate that smell," he screamed at Cabana. "Sweet, sickly smell."

Cabana, who needed education from no one about sickly smells, shrugged his shoulders and said they had other things to discuss. Johnson calmed down, and they spent twenty minutes alone together in the cell. For his last meal, Johnson wanted shrimp. He also wanted to wear regular prison jeans to the chamber, instead of the red death row jumpsuits. Cabana granted the request; he also offered Johnson the opportunity to stay with his family until late in the evening of his execution, which was unheard of. He understood the policy of ending the final visit early: you wanted at all costs to avoid the moment's emotionality. Johnson seemed to understand this better than he did, and said he wasn't sure he wanted his family there until the final moments.

That brought Cabana to a question he dreaded. He was required by law to ask Johnson if he wanted to be sedated. Mississippi offered a last-minute injection of Valium. It seemed like the kind of question that would bring the reality of the execution home to Johnson. Until the ammonia, he had seemed almost oblivious to what was going on. His grandmother had noticed it, too. She said Johnson seemed "happy and gay" every time they talked. "I just don't know if the boy sees and hears what position he is in," she said.

But that was Johnson in a nutshell. He knew a lot. He just never let you know that he did. Behind the thick glasses, Cabana could see the thought process about Valium work itself out in Johnson's eyes. As he'd feared, the subject had hit hard. Then he saw the decision being made. Johnson blinked slowly, like a curtain coming down. "I want a clear mind when you walk me in there," he said. "Will you be needing one for yourself?"

The irony and directness of the question made Cabana recoil. "No," he said, getting up from the inhuman concrete bed Johnson had slept on for nine years. At an American Correctional Association convention, Cabana had heard stories about a tumbler of whiskey the warden in Florida had shared with a condemned man minutes before an electrocution. That was a good idea, he thought.

"Good. I want you to have a clear mind, too," Johnson said. "I want you to know exactly what you're doing when you execute me. I want you to remember every last detail, because I'm innocent, Mr. Cabana. I'm innocent."

Hocutt awoke early on May 19, 1987, with the same abruptness and vertigo he'd felt on September 1, 1983. He didn't believe Johnson was going to die, any more than he'd believed Jimmy Lee Gray would, but this time was different. Now he knew he could be wrong.

Cabana had tried to convince himself it would happen, so he would be prepared. The logistics had been well seen to, but the more important preparation, of steeling himself, was way behind schedule. He had driven to Jackson the day before, to accompany Governor Allain to morning mass. They'd had lunch with Bishop Houck and talked throughout about Johnson. Houck had made it clear that neither would be found morally culpable by the Church, but he also advised them he would be making a strong statement against the execution to the press at the governor's mansion at midnight of May 20. "I'll offer a prayer there for the two of you as well."

The night had passed as badly for Cabana as for Hocutt. Both men were on death row by 6 A.M. A long, strange day followed, with flurries of detail followed by endless pauses. The film crew was all over the row by 9 A.M., giving an urgency to what would otherwise really have been a typical day, but the sense of drama was far less than Cabana had expected.

In the early afternoon, however, a gnawing unease began to grow, and he phoned Father Art Kirwin, Parchman's Catholic chaplain, to ask if he could hold an afternoon mass at the prison chapel, an old, secular wood-frame building at the end of guard row. Cabana was surprised when Hocutt asked if he and his wife, Patty, could come along.

Father Kirwin read from the litany, and Cabana kept his head bowed. "I don't pretend to understand the mysteries of Christ," he prayed. "But I need help." Behind him the church door creaked open. An MSU officer entered, carrying a fax from the Supreme Court's death penalty clerk. The Court had denied Johnson's last-minute plea.

Cabana looked it over. For all his sadness, his protestations of understanding the reality of the moment, and even the butterflies in his stomach, he realized, looking at the transmission, that he really hadn't believed for a second that Edward Earl Johnson would die that night.

The word *closure* had entered the vocabulary of the death penalty, like some twelve-step buzzword, but the only finality thus far was the resignation in Johnson's eyes when he'd told Cabana he didn't want a sedative. It was totally different from the "closure" he'd seen when the family had finished their visit the night before. They were a difficult group to make sense of. The grandmother was clearly the significant person, but there were many others. Most perplexing was Johnson's mother, who had given him up at an early age and moved to Queens, New York, where she worked as a meter maid. Their relationship was tentative. She had come down, of course, and had visited every day along with Johnson's grandmother and the others, but she and her son had barely touched when the time came finally to say good-bye.

It was a moment that would make anyone think of his own mother. Cabana, an orphan, was surprised to find himself thinking about his birth mother rather than the woman who had raised him, whom he'd unflinchingly regarded as his mother. It had been months since he'd thought of his birth mother. He was taken from her too early to have any memory of her, or of New Jersey, where he was born. She was a heroin addict, in jail for armed robbery, apparently a difficult prisoner, too. In the state prison at Trenton, she had spent a lot of time in administrative segregation, or solitary. Cabana had always suspected his mother's criminality had some connection with the career he'd blundered into, with his various and ineluctable returns to maximum security facilities, perhaps even with his repulsive attraction to the chamber's big black chair. He had never before understood how primary the connection was.

"Art, I don't know if I can do this one," he told Father Kirwin at the altar when communion was offered,

The response from Kirwin was almost immediate. "You have a responsibility to Edward Earl. This is not about you."

"Of the tens of millions of people in this country," Cabana asked, "why has it fallen to me to kill this man?" He had meant to ask it aloud, but he no longer needed to. The call-and-response of the Mass was in his mind, and he asked the question literally of himself, as though he existed in two parts. He saw his mother, withering in adminstrative segregation in New Jersey in the late 1940s, and the answer came to him: "You were fated to be here."

Hocutt, sitting silently in the first row, could see Cabana's weakness as he knelt at the altar. He'd seen it throughout the rehearsals, tests, and the endless meetings that had been called. How many times had Cabana asked about emptying the chamber if something went wrong? How long would it take, if a last-minute stay came down? Was there any chance of his being in there and your not being able to get him out? Cabana was a stickler for details, but this was like a child asking, Why? over and over, to a question that has no answer. Something bad is going to happen, and you're going to be a part of it. It was like coming to a brick wall in the middle of a desert. You can't go around, under, over, and you can't turn back, because you're too far out. You have to go through it.

•   •   •

Fred Childs and Roger van Landingham strapped Johnson in at 12:01, taking care with the new head restraint. Cabana had told Childs to remind Johnson to breathe when he smelled the gas, and he saw Childs saying it as he patted Johnson's shoulder. He also saw the stunned-deer response. The curtains came up after the officers left the chamber. Cabana was shocked at how close the witnesses were. It was an embarrassing moment. Embarrassment was the last emotion he expected.

They had to wait for the phone call from the DA's office, which for some reason wasn't coming. There was apparently a problem in the chemical room as well. Some hard water had made it into the bath and the sulfuric acid was taking longer than usual to cook. "Let's go with it," Johnson said. "Let's go with it." Even with his head pinioned, Cabana could sense a rhythmic movement to Johnson's body, rocking to an internal tempo. He said other things, always in twos. "What's taking so long?" he finally asked. "What's taking so long?"

"It won't be long," said Cabana, stepping in for a half second. Then he stepped out, further embarrassed that he had just reassured a man that he would be getting killed without further delay.

"I guess no one's going to call. I guess no one's going to call."

It took another five minutes before the acid bath had fully cooked. Johnson sang a spiritual softly. The melody made Cabana, still standing in the doorway, "being there" for Johnson, picture his friend Bill Allain, sitting in his mansion's office in Jackson, staring at the white phone glowing in the dark. Johnson had said he wanted Cabana to remember every detail of the night, but the faces and objects in front of him were just blurring together. Cabana had thought it would be difficult, but it wasn't. It was hateful. Everything about it was. Cabana looked at the witnesses, and wished he'd kept the blinds shut. The Drew sheriff, who'd horned in through Charles Tate Rogers, was sitting next to the national TV guy who'd stormed Cabana's office to demand his interview with Johnson. Of course, he had drawn one of the witness numbers, and the prick wouldn't shut up, pointing and waving his hand at this and that. What a pair, Cabana thought.

The phone rang. Sonny White told Cabana it was all clear. He stepped in and started reading. "By the sovereign authority of the State Supreme Court of Mississippi . . ."

Why am I reading? he wondered. This kid's been expecting this for nine years. I'm not telling him anything he doesn't know. As in the church that afternoon, his mind went into call-and-response, every word evoking a question in his mind. And this is not by or for the sovereignty of any court or state, he thought. They're not here. Was it a rite for the witnesses? They could go fuck themselves. He looked at them, then back at the paper in his hand, and read on.

It wasn't for the witnesses either. They could be here or not be here. It didn't matter. No, all he was doing now was throwing one last little dart of humiliation: *Look at what you did. Now look what we're going to do.* Retribution was one thing. This was sadism.

Then there were no more words to read. He stepped back out, van Landingham stepped in, put the cyanide in the bowl, shut the door, turned the big wheel, and Charles Tate Rogers threw the lever. Cabana smelled almonds, and felt his heart pumping fast and hard. He wondered if everyone heard it, if everyone else smelled the almonds. He looked at the media witnesses beyond, scribbling as though their reactions and the details they were noting were of importance. He was struck by the posture of Johnson's appellate attorney Clive Stafford Smith, a famous abolitionist lawyer from England. Smith was the only one of the twenty-one witnesses who kept his head averted, clearly signifying that this execution was something not meant to be seen, and not meant to happen.

Johnson began to hyperventilate shortly after the yellow cloud reached his face. His head moved vaguely forward and back—there was a quarter inch of give to the headrest, which would have to be corrected—but otherwise, he sat inert. Cabana thought, numbly, that Edward Earl Johnson was essentially not in there. He had been a soulful young man, but the soul had left the body when he'd said no to the Valium, just as the life had been stripped out of the body for the nine years he'd sat on the row. That hyperventilation he was looking at was an autonomic response to a brain wave. The brain wave was moving through the shell of a man Cabana had read a death warrant to a minute earlier. The man had been singing spirituals a minute before that. Praying for two hours before that. Playing chess four hours earlier. It was a brain wave telling whatever muscle group controlled the pleural membrane that the lungs were to breathe, just like Major Childs had told him to. And that muscle group was over-obeying now, making the small muscles in his face contract in that weird way, then making them all slack, as though he were an astronaut going through G-force tests.

What a wonderful thing the mind is, Cabana thought, watching a wet spot appear at the crotch of the white prison jeans Johnson had asked to wear. The urine seeped down his thigh, and Cabana watched it, feeling an acid sadness creeping into the coldness he had wrapped himself in just before midnight. That urine was the saddest thing he had ever seen. He felt so sad he couldn't help that young man, so sad he couldn't stop him from urinating on himself in front of all these horrible people. He looked at the doctor to his right, then back at Johnson. The fingers gripped the black armrest, and relaxed. Yellow spit foamed at the edges of his mouth, then stopped. Cabana looked at the doctor again, then at the EKG's green line, which was all but flat.

Eight minutes later, Johnson's chin rose and an unearthly, guttural sound, something between gas escaping and a groan played backward on a turntable, reverberated in the chamber. Its volume made Cabana's neck snap to the doctor on his right. Even in Vietnam, watching death play its various ways out of human bodies, he hadn't heard anything remotely like it. It wasn't death. Could the chamber be emptied? He looked to his left at Hocutt, and couldn't remember what Hocutt had said all those times about evacuating the chamber.

"Don't worry," said the doctor to his right. "He's dead."

Cabana hadn't asked. He certainly had no intention of asking why he shouldn't worry. Johnson's EKG wasn't flat by any means. It was higher than when he had last looked; in fact, it was very high indeed. Cabana looked at Presley and remembered him shuddering after the rabbits had been tested. "Jesus Christ," he suddenly thought, feeling like he was waking up to a nightmare. "I've got a young man in prime physical condition in there." The muscles in Johnson's face were rippling and his eyes were wide open. The ripples ran down the arms and legs, and Cabana knew he was looking at a body in seizure, and he felt the panic Johnson would certainly be feeling, if indeed anything was being felt in there.

It took fourteen minutes for the EKG to flatten. By the end, Cabana was well on his way to regarding what had happened not as a warden-insider running the thing but as an outsider, an abolitionist. Jimmy Lee's death had taken eleven minutes. If that had been cruel and unusual, what was this? A dental headrest from the 1940s didn't make for humaneness.

Cabana oversaw the dusting-down, the coroner's jury, and the removal of the body to a hearse from a Leake County funeral home, which would dress Johnson in a midnight-blue suit they had donated and bury him in a blue coffin with a mirrored inner lid, picked out by his mother from a catalog before she'd come down from New York. At the 2 A.M. press conference at the administration building he answered questions and stared blankly at the sentiments offered by the press. Then the night was over. It was a hot one, and the air-conditioning was long since off. The hot blasts of late May were the cruelest, particularly for a northerner like Cabana, who after fifteen years down south still hadn't adjusted to the weather. He stayed in the shower for what seemed like hours, scrubbing himself. Like many early risers, he was somewhat obsessive about feeling clean before getting to bed. That night, he just couldn't seem to stop sweating and get clean enough for bed.

Hocutt went to the Magnolia Savings Bank the next morning with his check from Leake County. *Pay to the Order of Donald Hocutt*, it read, *Two Hundred Fifty and 00/100 Dollars*. The note on the bottom left corner made the teller's jaw drop: *For the Execution of Edward Earl Johnson*. She counted the bills out without a word, but Hocutt knew she'd be showing the check to everyone when he left. But she didn't wait, and he stood in place with the bills in his hand while everyone behind the cage stared at him. When he went in with his check for killing Connie Ray Evans, seven weeks later, he'd make sure he got the same teller.

Charles Tate Rogers received his executioner's certificate in an eight-by-eleven-inch manila envelope from the secretary of state a week later. It was clean and formal, with a nice gold star on the top right corner that made it very attractive. He told Cabana he was going to frame it.

The BBC documentary, *Fourteen Days in May,* was released six months later, and Cabana was flooded with phone calls about the rabbits. He showed up for work one morning and was told that an irate Englishwoman was on the line.

"What bothers me," she said, "more than the disregard for life demonstrated in slaughtering defenseless animals, was the poor taste you showed with black rabbits. Didn't you see the connotations? In a prison in the American South?"

Cabana cleared his throat, and modulated up to his thickest drawl. "I appreciate your call, ma'am, I do," he said. "And you'll never know just how hard it was to find them black bunnies. Seems like all we got in our cotton fields down here's them tiny white ones."

But at 8 A.M. on the morning after Johnson's execution, he sat staring at a Styrofoam cup of coffee on his desk, wondering: How do you get to work after that? The phone rang. It was Morris Thigpen, calling from Alabama. Thigpen knew Cabana well—knew that he started his days early, that he wouldn't alter the ritual, even on a day like this, and that he'd be a mess and need a friend to talk to. "Something, huh?" said Thigpen.

"Does this get any easier?"

"I guess you'll see."

Unlike Johnson, Connie Ray Evans wanted to know everything about the chamber. "I don't want any surprises," he said. "Particularly at the last minute. Tell me about the last minute." Cabana told him how the phone call would come from Jackson, giving the go-ahead, and how he would then read the warrant. He said it probably wouldn't hurt, but that Evans had to breathe; he also told him about the headrest. Evans wanted to know why he'd be dying at midnight. He'd always figured it was so people wouldn't know, but that didn't make sense now that they'd been through one. Everyone was wide awake when Edward Earl Johnson died.

"Nope, nothing as elaborate as that as far as Mississippi is concerned," Cabana explained. "It's just a legal thing." If the chamber malfunctioned, if the lawyers got a restraining order or some other encumbrance that lasted only a few hours, the state had a full twenty-four hours without having to go back to the state supreme court for a new date.

That made Evans snort with disgust. Soon, everything Cabana said made him furious. Where warmth and trust had passed between them, there wasn't a subject now that didn't make them fight. Evans was insistent, childish, bitter, and amateurishly sarcastic. Cabana was like a bad parent who knew to count to ten before responding but for some reason just would not do it.

Evans was particularly furious about being asked for his last meal request. He finally settled on an omelette and fried okra—he was a vegetarian—and gave in when Cabana suggested he have a piece of chocolate cake. He said he didn't want his parents visiting late, like Johnson's had, then exploded when Cabana said *he* wasn't going to be the one to tell them. Evans had to.

"Do your damn job!" he yelled. "Turn them away if they come!"

"Do *your* damn job, and tell them not to come yourself," Cabana yelled back, wondering why he was being so ungracious. What was so hard about his telling Evans's mother to leave early?

The sounds of the chamber being tested enraged Evans. "You guys really think you're going to kill me," he shouted when Cabana came to visit the morning after the first rabbits. "My lawyers are still working." He wouldn't stop about how sadistic their testing was, and finally Cabana gave him the details of Gallego's execution—how a botched job could lead to an hour of torture. It was a good explanation for why they had to test the chamber, but it was no palliative. Evans was coiled in a fetal position when Cabana left the cell, and deep sobs echoed down the quiet tier, muffled only by a pillow Evans had thrown over his head. Cabana was on the verge of tears too. Evans had asked about soiling his pants in the chamber. Someone on the row had told him Edward Earl Johnson had wet himself. Evans's embarrassment, asking the question, felt like a sledgehammer.

The tests were not going well. Hocutt suffered some nasty burns when a bottle of caustic soda exploded in the sink during a practice run. It was a concentrate, but mislabeled. He was saved permanent disfiguration by his gas mask, but the explosion blew a three-foot hole in the ceiling of the chemical room. Then the first live test had brought them all within seconds of death. Van Landingham, again assigned the task, had knelt down to put the pound of crystals in the bowl after the rabbits had been put in the chair. Suddenly, he bolted up and yelled for everyone to run as he got the door closed. The crystals had gone straight down the well. Despite the checklist, no one had noticed the executioner's lever was still down. There was no doubt. Connie Ray Evans's execution was going to be a monster.

It got almost unbearable Sunday night, when Evans's final visit with his mother came to an end. Cabana watched the final embrace from the unit sergeant's doorway, standing with Chaplain Padgett and a young guard from the death watch, who was greatly moved. "I'd have to be ripped away from a moment like that," he said.

From ten feet off, Cabana heard Evans say, "Don't cry, Mama. I'll always love you." He let her go, then nodded to Cabana and toward the door to be taken back. What a man, thought Cabana. Evans had been right, too, wanting to cut this visit short. Why extend such a moment? The steel door opened, Cabana watched Evans shuffle his feet as he forced himself back to C Tier for the last time, then he looked down at his own shoes.

Seconds later, a woman's pair of shoes was standing opposite. "Do you have children?" Evans's mother was asking him.

Cabana looked up and his mouth opened. He had no idea what to say, and just nodded his head.

"Please, sir, don't kill my baby. Don't take my child away."

Cabana couldn't remember what the movie wardens said at this point. "I'm sorry"? He looked at Deputy Warden Presley, the young death watch guard, and at Chaplain Padgett, then thought about his mother up in Massachusetts, the one who'd raised him. "I'm sorry," he said.

With Cabana's approval, Connie Ray Evans spent his final afternoon reading from the Bible in the cell of another condemned man, Leo Edwards, who would be asphyxiated two years later. Ten years after Edwards's death, Cabana would remark on the fact that of the dozens of men who'd been on the row the day that Connie Ray Evans died, almost half of them slated for the chamber, Leo Edwards would be the only other man to die. Somehow, Connie Ray Evans had known that, and Edwards had known it too. "Connie Ray and Edward Earl really appreciated what you did for them," he'd tell Cabana months after Evans's death. "I hope you'll do the same when my time comes." Cabana didn't like Edwards. He was cocky, and his crime was repulsive. He and another man had escaped from Louisiana's Angola State Prison and gone off on a crime spree; after a few robberies they'd started killing at random. Edwards was condemned for the murder of an old man in a tiny convenience store in the middle of nowhere. He just shot him in the back of the head for no reason, then went back into the car boasting about how he'd blasted the man inside.

Cabana tried to tell Leo Edwards he'd wind up one of those old gray prisoners with a long beard, but Edwards shot him down. "No, sir," he said. "I'm afraid not." These men knew, and somehow found each other, even if their cell distances made that physically impossible. Connie Ray Evans had been Edward Earl Johnson's best friend on the row.

Hocutt was sitting in the sergeant's office of the old MSU when the fax on Evans came in from the Supreme Court. The denial surprised him. It came four hours later than Johnson's had. Somehow, he'd thought that meant it would amount to a stay. Cabana was eating dinner at home when Presley brought the copy of the fax over, nodding hello to Cabana's wife, Miriam, and the kids as he handed it to Cabana. "That's it," Cabana said, loosening his tie. "It's all over."

"He wants to know if you're coming back," Presley said, turning to go. "He wants to see you."

"I've caused so much pain to so many people," Evans said the moment Cabana stepped through the cell door. It had taken three hours to get down to the MSU—Cabana told himself he wanted to give the lawyer time to explain the decision—and Evans had a lot to say to him. What he'd been searching for in six years on the row was clear in his mind. He and Cabana had discussed redemption many times over the years, and both men felt the Bible was clear. The key was in your heart. He still didn't know why he had shot that man in the convenience store, but God did. He also knew that God had forgiven him, over and over, but in the time he had left, he wanted to see if he could forgive himself. Heaven was not an abstraction for Connie Ray Evans. He wanted to go, believed he was going, but he wanted to forgive himself before he did, and he seemed to want Cabana—as his jailer, as his friend—to open the lock to that self-forgiveness.

It was Evans's crime, though, his death, his redemption, and Cabana really had nothing to say. In his mind, however, he could feel the aura of the place—the "charisma of the condemned," the mystique of the MSU—suddenly lifting. Men like Connie Ray Evans—by chance, compulsion, design, fate, luck, whatever—had murdered other men, in deed rather than thought. But it didn't make him any different from his species, all murderers at heart. Anyone who'd been at war knew that. And anyone who had executed Edward Earl Johnson could tell you that too. It wasn't any great secret. But how do you get into another man's heart and tell him? How do you say, "That's okay, son. It could've happened some altogether different way. It just didn't. It's random. You're forgiven."

Cabana lit a cigarette and shrugged his shoulders. "When you get up there," he said, "put in a good word for me."

As he had many times over the past three years, Evans told Cabana, "Those cigarettes are going to kill you one day."

At 12:01, Evans looked around the chamber as they strapped him in—at the crucifix on the wall, the clock to his right, the doctors behind the window to his left, at Charles Tate Rogers to the right, at the open door, and finally at Cabana, standing at the mouth. Cabana followed Connie Ray's eyes, and felt there was something horribly wrong. "Of course there's something horribly wrong," he told himself. "I'm killing a friend of mine tonight." Then he realized that Childs and van Landingham had forgotten to draw the black drapes. The witnesses were seeing everything, all the preparations. Worse yet, he could see them. Stepping away from the chamber mouth, he asked Chaplain Padgett to spend Evans's last minutes with him. "I can't do this," he said.

A step outside the door, however, he turned and looked at Evans. He had been so brave and explicit, asking about this machine that would kill him, but his look was exactly the same as Johnson's: hypersensitive to the slightest details of the chamber but numbed, impassive. Cabana understood. If it was him, he'd thought with Johnson, he would have willed himself into a coma, or they would have needed four men to carry him in, biting, screaming, clawing, spitting, begging, crying, anything imaginable to fight or melt them. Like Johnson, Evans had turned off. This time Cabana had too. The eyes of all the witnesses were on him, but he wasn't there.

Not for them, at least. He was there for Connie Ray Evans, in a way he really hadn't been for Edward Earl Johnson. As their gazes locked, Cabana saw that Evans felt the same way. He no longer looked at the chamber, just at Cabana, with an intentness that was both strange and familiar.

The red phone rang. Cabana remembered telling Evans it would happen this way. He looked at the phone, and back at Evans, who was still looking at him. Their eyes locked again, and Cabana felt them falling into trance as the phone sounded again. Evans's eyes were huge and full of trust and pleading and resignation. It was impossible to tear away from them. "Mr. Cabana," Presley said, holding out the red phone.

Cabana felt his life written in that stare. It was Evans's life, but that made no difference. It was life about to be lost. In the classic image, he knew, a part of the executioner dies with his prisoner. Now it was palpable. Cabana felt a part of his life slip away.

Presley said his name again. "I'm getting out of here," Cabana thought, suddenly aware that his heart was racing. He remembered how it had beat the same way for Johnson. He wondered why he would inflict that horrible drumming on himself twice.

He finally took a step to his right for the telephone. Sonny's voice, telling him to go ahead, was filled with sadness. Cabana hadn't broken eye contact with Evans, and it wasn't necessary to say a word or shake his head. The only thing left was to read the paper.

Last time, he'd despised the warrant. Now he hated the clipped sound of his voice, his fake southern accent, until the final words—"Do you wish to make a final statement?"—came out, so dispassionately.

"I do have something to say," Evans told him. His voice was calm and intimate. "But I want to say it privately to you."

Cabana took a step in, feeling the horrible repulsion of the black chair as he approached, and his warmth toward the man sitting in it. The chair was so high he didn't have to bend to get his ear to Evans's mouth. "I'm embarrassed," Evans said after thanking Cabana for his friendship. Somehow, Cabana knew Evans wanted to offer absolution, as he'd asked for Cabana's. Didn't the condemned always absolve the warden?

This one was a dagger. "From one Christian to another," he said, "I love you. You can bet I'm going to tell the Man how good you are." Cabana reached for Evans's arm and squeezed it. Then he stepped out of the chamber, van Landingham darted in with the cyanide, and Cabana found himself next to Charles Tate Rogers as the chamber door whooshed closed and the wheel was turned. Presley said everything was ready, and Cabana looked at the chemical room door, where Hocutt and Parker stood with their arms folded. Cabana could tell Hocutt was wondering why he'd gotten so involved, and he knew Hocutt was right. There was no point in it.

Rogers didn't throw the lever. He was looking at Cabana. "Do it," Cabana said, to no one in particular as he looked back at Evans. His mouth formed the words *I love you, too.* Rogers looked at him quizzically. Apparently, Cabana hadn't said "Do it" out loud.

"Do it," he screamed. The lever dropped with its grating noise, and Rogers shouted the time to Ronnie Fulcher, whom Hocutt had asked to be the log keeper. Evans breathed quickly several times, then held his breath while the cloud hit his face. His fingers squeezed the edge of the chair, his lips pursed to slits, and his eyes darkened with the effort. Whatever part of his brain had turned off was back on now, fighting. The thin muscles of his neck stood out.

"Jesus Christ," Cabana screamed, startling everyone. "Make him breathe!" Evans's lips opened for three quick, involuntary gasps, and he was on his way. It took fifteen minutes for his EKG to flatten, and the death seemed to go a lot harder than Johnson's: more seething, eye rolling, and horrible rippling of the muscles, an even louder gaseous groan halfway through. A compulsive exerciser and health food fanatic, Evans was in superb physical condition.

Hocutt had decided Evans would be hosed down while he was inside the chamber, rather than on the table in the last night cell. Cabana watched Sergeant Cooke and an assistant do the hosing, then he joined Hocutt in signing the death certificate on the coroner's jury. It was 1 A.M. before he left the death house for the press conference in the administration building.

Miriam had spent the night packing the car and was waiting at the wheel outside the administration building. They were headed for a few weeks' vacation with the kids later that morning, their first since coming to Parchman. "No more," Cabana said, shaking his head to Miriam before walking in to do the press conference. "I don't want to do this anymore."

He was talking about Connie Ray Evans and executions, but as he said it, he realized he was through with prisons. Something had broken inside him, and he'd never be able to put it back together. Eight months later, he would leave Parchman to run a medium security facility in southern Mississippi. Three years later, he'd be out of the DOC entirely, getting his doctorate and teaching criminology at Hattiesburg, traveling across the country as a leading abolitionist: consulting, speaking, and testifying as an expert witness. In twenty years of prison work, Cabana had never questioned his own authority, his own correctness. It had always seemed true by definition. Evans's final moments had stripped him of that claim, or at least the immediacy of being able to make it. Without it, there was no way of really running inmates. He would have a title but no substance, and there was no faking it in his line of work.

At the press conference, everyone wanted to know what Evans had whispered in the chamber. He refused to say. After a long silence, they asked about his mood. "As well as can be expected," he said, "given the circumstances." With nothing left, they asked about Evans's last meal, and he gave them the menu, adding, "He remarked that he was particularly appreciative of the chocolate cake." The last reporter left at 2 A.M., and Cabana and Dwight Presley locked the building up behind them. The moon was high and brilliant, and the two men looked at it before Cabana opened the passenger door and got in with Miriam. Presley asked him what he was going to do.

"I think I'm going to call my mother," he said.

CHAPTER EIGHT

"HAVE YOU EVER THOUGHT ABOUT CHANGING JOBS?": THE EXECUTION OF LEO EDWARDS

TO HOCUTT, RUNNING INMATES, IF YOU DID IT FROM THE HEART, was like being a rubber band that someone kept twisting. Eventually, you snapped. Six months after Don Cabana left Parchman, a broken man, something inside Hocutt snapped. The rupture in his psyche would take six years to play out to the bitter end, and he hid it well at first. Outside the house, he seemed the same old velvet fist in an iron glove: hard drinking, hard living, hard on convicts, still full of extreme practical jokes. To himself and to his family, however, the change was clear. Where he had always been quick to get angry, express it, and move on, now he seethed for days on end. Weeks passed in which he simply could not stand to see another convict, and he spent his shifts behind the GMC's wheel, driving in circles for hundreds of miles.

He overate and ballooned to 300 pounds by the winter of 1988, when ice storms hit the delta and closed everything down. It was beautiful and surreal to look at—trees, houses, and cars contoured in thick, opaque ice slabs—but at night you'd hear the tree branches, telephone poles, and electrical wires groaning and coming down under the weight of the ice, and it was spooky.

Hocutt had a major case of the creeps all winter, just stepping out the door onto the prison grounds. In the convicts' language, he was doing hard time. His shifts had always been a kind of salvation, and his progress up the ranks, however torturously achieved, was a rock-solid base for his ego. Now, each day seemed to take a piece away from him, and he didn't have the will or the means to stop it.

As Leo Edwards's date with the chamber, June 21, 1989, approached, Hocutt was given a chance to disengage from the stress of running inmates: a deputy wardenship, offered by Steve Puckett, the new superintendent. Hocutt turned it down. He liked leadership, tolerated his share of administrative paperwork, and had even developed a flair for working with people he hated—a huge personal milestone. Wardencies were political positions, however, and after fourteen years as a Parchmanite, working well below the radar of the DOC and the politicians in Jackson, Hocutt knew that he could never be answerable to anyone outside these walls. Certainly not the way he was feeling. The capacity to be inside them was all but gone.

He and Patty decided to get off the prison grounds at least, and started house hunting. They found a big, solid ranch by the banks of the Tallahatchie River, thirty miles down 49W in Greenwood, bought furniture, and moved in early June. Thirteen years had passed since Hocutt had moved into Parchman. A few hours after the movers had put the new oversize plaid couch on the white shag carpet of the living room at the ranch's north end, Hocutt was on it, looking out the dining room window at cormorants foraging on the levee at twilight, and at the sun setting behind them. Then he lay down, crossed his arms and ankles, and stared up at the white stucco ceiling. Eight hours later, he awoke in the same position, with the sun shining through the living room window. He dressed in his uniform and drove to work.

Every night after dinner, he'd lie down on that couch, sigh, look up at the ceiling in a rapture of exhaustion, watch the way his life

had played out, then awaken the next morning with the ceiling lit
by the sun. He never believed he'd fallen asleep. He'd just rub his
eyes and go to work. Though he now had a big house to spread out
in, and the world at large to explore, he was acting like any prisoner freed after long captivity: confining himself to the small spaces
he'd been restricted to.

From thirty miles away, he could at least see why Parchman had
become so joyless. With farming slowing down again, never to
resume at the pace set during Cabana's four years, the prison was
following its old entropy, down to paperwork, codes, and half-
hearted claims to reform—all the bullshit and buzzwords Cabana
had put a halt to. Hocutt believed in rehabilitation but felt it had to
come from within, and the inmate had to really want it and understand it was a privilege, not a right. Applied like a Band-Aid, with
procedures from some Jackson office, it wasn't reform at all, just a
passing of the buck, at best. More likely, it just meant some politician or bureaucrat was fighting for turf he'd never see again.

To run inmates you had to get into their lives, with authority.
That didn't necessarily mean the short course, but if it didn't begin
with letting them know they were at Parchman for punishment and
to work for the state they had harmed, there would be no end to
the troubles. Cabana had been right—reform came at the end of a
hoe. Parchman was a farm in the floodplain, which meant that you
grew cotton. In a good year they had been hard pressed to break
even, but the dividends from a penological point of view were obvious and had accrued quickly. Two years after Cabana left, half the
acreage would be fallow, and half of what still was farmed was contracted out to the free world. Inmates raised a modicum of cotton,
but it was gentleman farming, which is a delta oxymoron. Instead
of working the fields, inmates drove trucks that sprayed bulbs with
defoliant to make the petals open, then big green machines with
twenty-foot retractable arms did the picking. The inmates still
loaded the big trucks and bagged the cotton, but that had always
been a job for trustees with ten to twenty years of good time.

There was no bucking the flow, of course. This time, however,

Hocutt couldn't bring himself to fight for domain: for his ideas in the morning meetings, over camps, for his share of Superintendent Steve Puckett's ear. That included the executioner's job, which came open when the Mississippi Highway Patrol, which hadn't cared for the publicity of executions, announced that Charles Tate Rogers would no longer serve "at the pleasure of the governor." Everyone expected the job to fall to Hocutt, but Fred Childs got the commission for Leo Edwards. It was a slap in the face, and cost Hocutt the executioner's extra $250, but he took the decision in stride. In the two years since Edward Earl Johnson and Connie Ray Evans had gone, death row had been moved to Unit 27, Hocutt had relocated his personal office to another building, and he was largely disengaged from the old MSU, except for his work with the chamber. But Leo Edwards would go as his two friends had, with Hocutt running things from the chemical room. As it had with Cabana's tenure, it would follow his protocol—his meetings, practices, and orders.

He was an "executioner without portfolio," but everyone in the delta knew he was the man. In Greenwood, people told him "Keep up the good work" as they passed—on the street in town, in the aisles at church, or in Wal-Mart. A salesman at the Caterpillar shop, who was under the impression that Hocutt executed people on a weekly basis, screamed when Hocutt grabbed his arm and gave it a squeeze after the man quoted him a high price on a backhoe. A car salesman named Jimmy Gray kept asking for top dollar, though Hocutt had told him he hated haggling. "Listen, you polyester-wearing, chicken-eating motherfucker," Hocutt finally told him, "I smother people with your name." Gray went slack-jawed and they settled shortly after. A man Hocutt met at a party argued that the gas chamber hadn't deterred a single murder, then told Hocutt he didn't even want to talk to a killer like him; other people told Hocutt they wanted to throw the lever themselves. When he assured them that they really wouldn't—at least not more than once—a surprising number said they just wanted to hold on to him while he threw it.

He felt his death house status most keenly inside the prison. Cruising the grounds one afternoon, he got a call on his radio about an inmate who had squared off with a young officer in a cotton field a mile away. By the time Hocutt arrived, the officer had already drawn his gun and fired into the ground, but the inmate had backed the officer up with a hoe and was advancing steadily. Hocutt stepped out of the GMC with his .45 raised, resolved to shoot if needed. The convict turned in his direction and raised the hoe higher. Hocutt leveled the .45 down with both hands on the grip, until the bridge of the man's nose was at his sight guard: it was like target practice, and he felt no different than he did on the range. He got his finger on the trigger and breathed in. Two inmates dove in, brought the man with the hoe to the ground, picked him up, and carried him to the back of Hocutt's truck to be bound hand and foot.

Hocutt was shocked. "Why did y'all save me?" he asked. "Tackling him like that?" He didn't know either of the two who had brought the man to the ground. Saving an officer, a ranking one at that, could get you killed at Parchman.

"We were saving *him*," said one. "You put a nigger in a tank and smother him, you'd shoot one down in a cotton field, for sure."

Hocutt smiled. The next time he met some statistic-toting abolitionist arguing that the death penalty hadn't deterred a single murder, he'd have a statistic of his own.

In the second week of June, Hocutt had Leo Edwards brought from Unit 32 to the last cell on C Tier, and put him on a twenty-four-hour death watch. A flurry of visitors checked their car keys at the double-tiered gate and came in—everyone from abolitionist lawyers like Clive Stafford Smith, who had fought to keep Edward Earl Johnson alive, to Leo Edwards's daughter, Latareaha, whom he hadn't seen in the nine years he'd been condemned. But the drama of his final days was considerably lessened, now that the row was in a different building. Edwards had come within fourteen hours of the chamber eight months earlier, when the condemned were still in the old MSU. He'd been saved by a rare eleventh-hour Supreme Court stay, and the tension that had seized the building in the final hours, and its release when Edwards was stayed, had brought the reality of the penalty back to Hocutt like a splash of cold water.

As this new date approached, however, Hocutt realized that the drama and monumentality he'd been living with for so long was all but gone. Only three men had died in the decade he'd overseen the MSU, but there had been more than two dozen death watches, and the aura of death had hung over the building like the smell of burned toast. Now it seemed the entire country had fallen asleep, as far as the penalty was concerned. There were exceptions, of course: The lethal injection of George "Tiny" Mercer caught the national eye in the first week of 1989 (it was Missouri's first execution in almost thirty years); eighteen days later, coverage of Ted Bundy's long-awaited electrocution in Florida got type sizes and column lengths equal to that week's two other big events: George Bush's inauguration and the Super Bowl. But executions had otherwise become rote procedures. The lethal-injection gurney had seen to that.

There was some controversy over Edwards's June 21 date, stirred by charges of "southern justice" leveled by Clive Stafford Smith. Edwards had claimed since 1983 that his prosecutors had systematically struck every potential black juror from the panel that condemned him in 1980; Smith, a prolific antipenalty propagandist as well as an appellate lawyer, had done much to get that story out to the press. In truth, Edwards had no other claim to raise. There was no question of his guilt; he was a career criminal; his three murders were unprovoked and undeniably cold-blooded; and the one for which he'd received his capital conviction was of a black storekeeper, so even the proportionality bias cited in *McCleskey* could not be raised. Instead, his execution date was lent symbolic poignancy by the coincidence that it was the twenty-fifth anniversary of the murders of civil rights activists Michael Schwerner, Ben Chaney, and Andrew Goodman, and in Jackson and in Philadelphia, Mississippi, a hundred miles east, the date would be marked by memorials and demonstrations.

Hocutt detailed security teams outside the prison for what he considered to be the inevitable mass gathering and candlelight vigil, but as Edwards's final hours approached, it was clear this execution had failed to become a national issue. The few news stories about opposition to the asphyxiation in the day's papers concerned an abolitionist collective in Cambridge, England, that had taken up Edwards's cause. Less than a dozen protestors showed up outside the gates that evening.

Edwards asked to be sedated at the last minute and received a Valium injection at 11:30. Thirty-four minutes later, he was half carried, half dragged to the chamber, and his body was rubbery and his eyes drooping as van Landingham fitted his head inside the headrest and strapped down the arms and legs. Childs darted in, poured the cyanide below the chair, turned the huge wheel to bolt the chamber door, waited for the gauge to go up, and threw the lever. From the chemical room door Hocutt watched the gas rise. Edwards's somnolent body sagged heavily against the straps after his first few breaths, but there was no sign of struggle.

Five minutes into the asphyxiation, however, the gas appeared to rouse him. His eyes opened, he stared straight ahead, and he cried out twice before going into involuntary spasms and paroxysms. A full fourteen minutes passed before the doctors could declare death. Hocutt still couldn't believe the stress involved, particularly when Edwards seemed to come back halfway through. This time, however, he felt like he had been drugged himself, and he watched more than felt his own alarm and panic when Edwards's body went through its strange movements inside the chamber. Hocutt, who believed those movements were made insensate, felt as though his own natural empathy and instinct had become autonomic.

He told himself he didn't care for Edwards, and that he'd gone cold—that the shock of Bruce's war whoop and Jimmy Lee Gray's head-banging had anesthetized him to everything that came after. Cabana, he also realized, had done a man's job with Edward Earl Johnson and Connie Ray Evans, taking the entire onus on himself. To Hocutt's mind, that had been an insane burden, getting emotionally involved with the condemned at the moment of their deaths. The mere fact they were being executed already aggrandized them in their executioners' eyes, which was a natural response. These were reduced men; in their extreme confinement, barred from any self-reliance, dependent on outsiders for every stitch, morsel, and ability to clean themselves, they were the purest wards of the state. Taking on a paternal role—confessor, sympathizer, facilitator—as Cabana had, particularly at the moment he had killed these ultimate wards, all but sanctified them with sympathy.

Hocutt hadn't planned to stay for the cleanup after the coroner's jury, but he didn't leave the death house until the ambulance with Edwards's body had left the MSU. It was after 3 A.M. when he took his place on the couch back at home in Greenwood, and Edwards didn't cross his mind as he drifted off. There were too many other things to think about, and after three weeks at the ranch, Hocutt knew how soon morning would come.

He rose before the first light cut through the living room window and walked quickly in his white-stockinged feet down the hall into the kitchen. Opening the walnut cabinets above the kitchen sink, he riffled through bags of rice, paper towels, and cans of soup and vegetables. There was a mist on the river that he could see out the kitchen window, and it looked ghostly in the ambient predawn light. It was the deadest hour, when night was over but morning hadn't come. He was surprised Leo wasn't in any of those cabinets. Something about the sink had made him sure he'd be in one of them.

He tried below the sink. There were only detergents, brushes, soap powders, and the garbage can. In the fridge, he found a six-pack of Diet Coke, ripped one can from the plastic, then drank half in a gulp and scratched his head. That left the broom closet, an obvious place, but Leo wasn't in there either. He had gotten through most of the wall of cabinets above the service counter between the kitchen and dining room when he saw the clock on the wall. It was half past five, almost time to go to work. Hocutt rubbed his eyes, and as always woke up convinced he hadn't gotten to sleep. He still had on his colonel's whites and the gray pants from the execution. Only his shoes were off.

He was reaching for one of the two cabinets he hadn't checked above the service counter when he realized what he was doing and laughed out loud. He thought to wake Patty and tell her whom he'd been looking for in their broom closet. She knew how much he hated Leo Edwards, how much he disliked vacuuming. It would take a dead man to get him to open that door. He shrugged his shoulders, laughed again, took a half step toward the bedroom, then reached for the Diet Coke. He sipped at the other half for another minute, looking behind the canned green beans and carrots in the last two cabinets, as if Leo had become a mouse and could hide behind cans. Maybe he wouldn't tell Patty. She wouldn't drop this one for weeks.

Neither could he. Night after night, he lay on the couch and drifted into a state that was like nothing he had ever known. He couldn't tell if he was sleeping or waking, dreaming or thinking; every few nights he'd get up and wander from room to room, looking behind furniture and in closets, then get back on the couch, wondering where Leo had escaped to. When Patty came in for morning coffee he'd be at the kitchen counter looking for Leo, staring straight ahead with his eyes bugged open.

The executioner's commission and certificate finally came to Hocutt in 1991, as a slew of dates approached. He thought the certificate was a silly formality, but he spent $100 getting it framed and put it on his office wall like a diploma, giving it pride of place behind his desk. It had a particularly strong impact on an impressionable young major named Thomas McDaniel, and Hocutt decided to milk the young man's gullibility. Out together in Hocutt's GMC one afternoon, they stopped outside Unit 27, where death row inmates exercised. Hocutt made a show of popping his eyes as they watched a four-on-four basketball game of condemned men from behind the unit's fence. "Five hundred times eight, carry the four," he said in his best stage whisper. "Three zeroes behind the comma . . ."

"Donald, what are you counting?" McDaniel asked.

"My fees," he said, making his eyes pop further. "Tom, that game alone's worth four thousand dollars to me." For weeks afterward, he heard about what a sick-hearted killer he was. Things had changed since Hocutt had started at Parchman, but the gossip mill ground as fresh as ever.

The dates for which he'd received the commission, however, passed without an execution. Every few months, his office would receive a fax from the Mississippi Supreme Court—another appeal had been denied—and he would begin to gear up at the old MSU. Then more faxes would come: a stay from the Fifth Circuit, denial from the Supreme Court; a stay from the Southern District Court of Appeals, denial from the Supreme Court; a stay from the state supreme court, denial from the Supreme Court, until finally the state supreme court clerk who had originally advised him of the date would fax him that the cycle of appeals was to begin again.

Crack cocaine had devastated Mississippi by 1990, and the new men on the row seemed younger each year—incorrigible in a way the Edward Earl Johnsons and Connie Ray Evanses had never been—and to Hocutt's mind that much more in need of the sentence. There were two guys he particularly wanted to execute. They had killed a family, sodomizing the daughter with a broom handle while her parents watched, shooting her in the head, then torturing and killing the parents in the same way. They were in a medium security camp now. Down the hall from them was a guy who had raped and killed a nurse and cut off her nipples, which he had in his pocket when apprehended. He was on the row less than three years before he was resentenced and sent to general population. That was unusually quick, but they all seemed to leave sooner or later. Over 80 percent of the men condemned to the chamber since 1977 were in population now. Two had made it off the farm altogether.

It made him think about the penalty as a political issue, which he'd never done before. It had always been a question of the state supreme court saying, Get ready, then some other court saying, Stand down. Now, he had a question of his own: Why am I not killing these people? He read articles and editorials debating the gurney versus the chamber and got so angry he had to laugh. Eight years had passed since the lethal-injection suite had been installed in Parchman's death house, but not one man convicted after July 1, 1984, the cut-off date, had made it as far as death watch. There were men in the legislature who actually argued that Mississippi *had* to keep the chamber simply because the FDA could intervene in a lethal injection if the condemned were allergic to sodium thiopental or Pavulon. It was ridiculous. What if they were allergic to hydrogen-cyanide? Soon there would be only five men facing the chamber, all with IQs below 70 or diagnoses of schizophrenia, and they were only on the row because of a legal glitch. Unlike most states, Mississippi didn't allow its juries to opt for lesser sentences for "diminished responsibility." But the higher courts had to.

That kind of state-by-state quirkiness antagonized Hocutt as much as it did the abolitionists. To his mind, the federal goverment should do the job, appeals should be limited to a single, well-funded round, the gallows should be the only method, and it should be as public as possible without inflicting harm on the families involved. Hanging was a good, violent death, and it made the point simply, quickly, and with a truly American resonance. People asking Hocutt about the chamber usually got his stock response—"So far, it's us four, them nothing"—even though he didn't think it was that funny. Now, he realized, it was dead wrong. It should really have been: "us four, them fifty-five," or however many men happened to be on the row at the time and not being executed. They made a laughingstock out of the penalty, and out of the State of Mississippi's will to justice.

As he'd realize during a consultancy three years later, in 1994, the *will* to do it was the only question. A man called from North Carolina's Central Prison, where the condemned had a choice between gas and the needle. An inmate's appeals had run out, Hocutt was told, and he had opted at the last minute for the chamber, knowing that it hadn't been used since 1961. It put the prison in an embarrassing position. There was no one there who knew how to run the thing. The man asked if Hocutt could draw up a protocol, test the chamber, perhaps run the execution. Hocutt wished it had been California or Arizona calling. He had no interest in a trip to Raleigh.

"There's a lot to know but little to do," he told the man. "What you really have to know is what *not* to do, and that's kill yourself or your witnesses." Running through the procedure step by step took less than an hour. A week later, Hocutt picked up the paper and read about the asphyxiation of David Lawson, North Carolina's first in over thirty years. *USA Today* gave it front-page coverage, more than they had given a month earlier to John Wayne Gacy's lethal injection. If North Carolina could kill despite such publicity without even really knowing how to do it, what was wrong with Mississippi? Even if they used the gurney, Hocutt could cut a man down for a vein in his sleep.

A new Parchman superintendent, Ed Hargett, was named in 1992. He offered Hocutt reassignment, overseeing district community work centers (CWCs), minimum security work facilities for nonviolent criminals, and Hocutt took the job. It was a relatively stress-free, bureaucratic position, half of it spent on long drives through the pretty northern parts of the state. At the office, he pushed budgets from one side of his desk to the other, read reports of health codes, and held conferences with the six captains below him about work assignments and inmate transfers.

It was the DOC's way of saying he had put in his time, the first rung of getting kicked upstairs. The rest was up to him. He could navigate between branches of corrections, put in for some wardency down the line, take a job in Jackson. Six months into it, however, he started getting ill. Hocutt, who hadn't used a sick day in seventeen years on the job, tried to shrug each ailment off, but the sheer number soon told on him. First came maturity-onset diabetes, requiring regular insulin shots; then diverticulitis; arthritis; partial deafness; and gout, which was easily the most painful. By the end of the year, he was losing weeks of work at a time. He had probably seen a half dozen doctors in his first forty years. By mid-1993, he had so many he couldn't keep their names straight.

The effect on his mind was worse, and more perplexing. There was simply no stress he could bear, and he exploded at the tiniest hitch. He had always looked on mornings as a fresh start, but he woke up now feeling hollow, empty, guilty, angry. At times he could shake it off and get to work and back, but more frequently he'd open his eyes to a black cloud that felt like a steel blanket of depression, or terror, or anger, and he couldn't make it out of bed. At times it sapped his strength; at others, it bound him in knots, and his muscles cramped. He'd ball up in bed and look for something inside himself to fight it off with, but there was no fighting this.

At Patty's urging, he agreed to see head doctors—in Greenwood, Greenville, Memphis. He told them he felt like the shittiest human being in the world, of feeling hounded, and he spoke of a "constant negative draw" that felt like it was sucking the life out of him. One psychopharmacologist gave him mood elevators; another told him he was bipolar and prescribed lithium. Another told him to get used to an episodic life until the Prozac or Paxson cured or stabilized the depression, or whatever it was. "Have you ever thought about changing jobs?" he was asked.

The final episode came when Superintendent Hargett, with whom Hocutt had begun to fight bitterly, asked him to come up to Parchman. They scheduled a Wednesday meeting, but Hocutt didn't make it. He'd backed his truck up on the thin spit of the levee above his house when he realized he couldn't drive. By noon Thursday, he was sitting with Patty in a clinical psychiatrist's office in Memphis, hearing what he knew already: he was through. The man told him he had no reservations about recommending a full medical discharge.

The DOC in Jackson and the people at Parchman didn't agree. Hocutt spent two years fighting it out. They told him he wasn't sick, just tired and irritable—who wasn't?—and their refusal to credit his illnesses made him livid. He'd run every goon squad in the place and sat in the dockets of civil courts over 500 times, hiding their dirty laundry for almost two decades, and now they seemed to be questioning his very essence. No one confronted him; they just didn't give him his discharge. Instead, he got commendations and medals: he was up to twenty-seven by the time the Mississippi Association of Professionals in Corrections made him Corrections Officer of the Year. But they didn't speak to him much at the conventions, and he could tell they were talking behind his back.

Up at Deer Camp one weekend, he wandered on to a hilltop on someone else's land and got into a terrible argument with the property owner. He knew he was in the wrong and apologized, but the man kept at him. Hocutt kept his counsel until he heard the words "you son of a bitch," then put his rifle down and let go for the first time in months, knocking the man to the bottom of the hill with one punch. Looking down, Hocutt wondered if that, simply, was what he'd been missing—good, regular opportunities for violence. On his rare appearances at the community work centers, where almost everyone put in good time and watched their language, Hocutt brought people to tears with his tirades. The effect of all this anger on his family was terrible.

A year into his fight with the state over disability, Hocutt began to feel his future as a wage earner was in jeopardy, and decided to try his hand at the catfish business. He went in with a friend, Terry Johnson, who handled everything from customers to bank loans, got himself discharged from the CWC, and started working full-time on catfish farms. It felt great to be sitting on a bulldozer again, making ponds out of old cotton fields, using lasers to lay grades. They built new farms, did maintenance on existing ones, and raised their own catfish; as their equity became sufficient to qualify them for any loan they needed, their equipment became increasingly sophisticated.

Then the winter of 1993 brought the worst ice storms the state had seen in fifty years. They wiped out fields and orchards from Greenville to Memphis, and shut down the entire north and western parts of the state for three weeks. At first, it seemed the storms would be good for business. There was limitless work repairing the damage to the grades, slopes, and levees of the ponds. But the ice also left the soils in a dreadful condition to drive heavy machinery through. On cold days, the gumbo froze and became so slick Hocutt had to drive the trucks in a kind of controlled skid. When it warmed, the soil became like quicksand that sucked tractors and bulldozers two or three feet down into the ooze. Then the rains came in February and stopped everything. Contracts went unfulfilled, plans went on hold, and the stress of watching helplessly as weeks of grading work came undone in the downpours became unbearable.

Patty and Terry persuaded him off the crew and into the office, set up in a double-wide trailer off a main country road. The change had come too late, however, and it quickly became clear that running the business was too stressful. Patty, who had gone to work as an LPN at Greenwood Hospital when they'd left Parchman, had to take over. His stepson, Chris, halfway through high school, had the Hocutt capacity for work, and started learning how to run machines, then small crews. In a few years, he would be able to take over the field work.

Hocutt bought a La-Z-Boy, a thirty-three-inch TV, got hooked up to enough satellites to see everything from Mexico to Japan, and spent his days flipping. He had 300 channels, but only history shows and NASCAR held his interest. Days went by with him on the La-Z-Boy, then months: half asleep, half watching. A new executioner was appointed by the governor, Major Robert Armstrong, whom Hocutt had trained.

Another nightmare began to plague him. He was gassing a man in the chamber, with two to go after him. There was no one around the death house to help, so he'd filled a cow trough with ice in an adjacent, barnlike room to the left, where the chemical room had been, and had put the other two men in the trough to wait their turn.

He'd wake from the dream feeling angry that there had been no one there to help him. The ice was the spooky part of it, because he couldn't figure it out. The significance of the three men in the dream, all former death row inmates who had gotten off, was easy to see, though none of them was a particularly heinous criminal. The ice chunks were huge. The men floated in the trough like boats waiting to hit an iceberg. Hocutt guessed it symbolized the end of the gas chamber, which was being put in deep storage all over the country. Legislators in North Carolina, which had consulted with him earlier that year, had moved to banish theirs after two officers conducting a 1994 asphyxiation had blundered and almost killed themselves. California's chamber was closed amid great fanfare, then Arizona's. Mississippi's was a relic, but the state seemed wedded to it. Month after month went by with sentences unfulfilled. Ten years after the introduction of the needle, the lethal-injection suite had gone unused, and the only men who'd made it as far as the death watch since *Gregg* were the ones who'd been slated for the gas.

In another dream, the sun was shining on Parchman's fields and the prison had gone back to farming cotton. The White Hats were on horseback again, taking the inmates out, deuced up or four wide, shovels over their shoulders, singing the work songs they'd always sung in the fields. Hocutt had a machine gun. He squared the men off and mowed them down, row after row. Then he woke up on the La-Z-Boy, rubbed his eyes, and started flipping from channel to channel.

AUGUST 10, 1998. The cell phone in Hocutt's silver Mercedes 510SL rings as we head up I-55 in a blinding rain for a weekend at Deer Camp. Patty drives and Donald sits in front with his arms folded over his chest, his bare, bone-white heels splayed on the dashboard. I eat Big Macs in back with their son Mark, who's fourteen now.

Hocutt lets the phone ring four times before picking it up with his brief "Yup?" He has phones in every car and most rooms in his house, but always answers as though he's been awoken at 2 A.M. and never stays on a second longer than he has to. This caller, John, gets an unusually warm reception. John has had his stomach stapled, so that it holds only an ounce of food at a time. After ten years of being overweight and five years of severe gastric problems, Hocutt has decided to get stapled as well, and has booked himself for surgery a month from now at the Little Rock, Arkansas, hospital that John recommends.

It's my last research trip in the delta. Freak atmospheric storms have closed airports in Memphis and Jackson and deluged northern and central Mississippi for three days running. The rain comes in ten-minute pockets as we drive through the miniclimate zones in the center of the state. Patty, in her nurse's whites, stays on the road even when visibility is less than twenty feet, pulling off only when the gushes make it impossible to tell the windshield wipers from the rain in front of her.

"Don, there is nothing to it," John says over the speakerphone. "I had one little bit of pain, from the medicine, and I felt light-headed for a very short while. Other than that, I really cannot recommend it highly enough." John has a strong accent but no twang or slang. His pronunciation is almost Edwardian, a manner of speech that marks some of the upwardly mobile down here—a minding of one's *P*s and *Q*s, lest a hint of country show through. Hocutt's diction has changed in a similar way over the years I've known him. He still makes jokes about being redneck trash, but they're all off Jeff Foxworthy records.

"So how much are you eating?" he asks John.

"Bits here and there. Peanut butter and crackers, in those little orange packs. You get used to that quickly—mini-soups, two-point-five-ounce cans of vegetables. One ounce at a time."

"Are you drinking juice?"

"Quite a bit. The little cartons, with the pop-in straws."

"You have any problems going to the bathroom?"

"None at all. Well, sure, some. But I'll tell you, Donald. A friend of mine went from a seventy-two-inch waist to a thirty-six. I'll take some potty problems for a month or two if I can pull that one off."

Hocutt has plans for the October NASCAR races in Talladega, Alabama. He went three years ago with Ronnie Fulcher, and each year since the family has taken a camper and stayed out on the track grounds for a week. Patty can't get off work this October, however, so Hocutt and Ronnie are going stag. John doesn't know if that's such a good idea.

"Donald, that's two months off. One month after the operation? You're going to need Patty with you, man. Now you and I both know you're not giving up Talladega, but you can get stapled any old time. I certainly would not relish being in that campground when nature calls. Whoops. Calling now." Click.

Hocutt shudders as he hangs up. "Man, I can't believe I'm going through with this."

"Dad, did they ever run around inside there?" Mark asks. We had been talking about Leo Edwards when the cell phone rang.

"Run around inside where, hon?"

"Inside the gas chamber."

"No, son, they couldn't run. They were strapped to a chair and their heads were pinned to a headrest."

"You strapped them to a chair and pinned their heads down?"

"Some other people did. They're called the strapdown team."

"Why'd they strap them to a chair and pin their heads down?"

"So the gas would hit their faces from below in the right way. And so they wouldn't bang their heads against a pole."

"And you watched it happen?"

"Sure I did."

"That's the death watch?"

"No, the death watch is when they're sitting in a cell by the chamber and you watch them twenty-four hours a day."

"You watched them just sit there for twenty-four hours?"

"Someone did. Sometimes for a whole week."

"Why'd you watch them so long?"

"So they wouldn't kill themselves, son."

"Could they see the gas chamber from the cell?"

"No, but they could hear me working on it from in there."

"Why were you working on it?"

"I was just making sure they died and I didn't."

"I still don't know why they don't just shoot them in the head."

Hocutt has been talking rather freely about executions on this trip. Normally, it's hard to get him to answer more than a few questions at a time, but the chamber is on his mind now. The state legislature finally banned it altogether a month ago, hoping to break the logjam on death row: It's been nine years since Leo Edwards died, and no one has been executed in Mississippi since. The other states that asphyxiated post-*Gregg*—Nevada, California, Arizona, and North Carolina—have been able to carry out sentences more easily with the needle since giving up on their chambers.

"I've tried to tell Mark and Chris about the gas chamber many times," Hocutt says, looking at me over his shoulder. "I want to tell more kids about it—get an easel, some pictures and diagrams of gas chambers, electric chairs, gallows, gurneys, before-and-after photos, and go to the grade schools around where we live and show the kids what this stuff is like. It's no deterrent if it's a secret. Little minds need to know. Look at that."

Hocutt points to a vast field overrun by kudzu. There is nothing visible for the vine to have grown on, but it's waist-high as far as the eye can see. Beyond the field is a forest of thirty-foot trees that has long been engulfed. The leaves, which thrive in this kind of rain, are beautiful and heart-shaped. When the vine extends for hundreds of acres at a time like this, it looks like some alien form of vegetative narcosis. I tell Hocutt about a kudzu recipe I'd seen in a folksy cookbook at the Memphis airport bookshop.

"People'll eat anything," he says. "If they're poor enough."

"You'd never eat kudzu, would you, Dad?" asks Mark.

"Son, I'll be eating it in a month. One ounce at a time."

Deer Camp is a few miles off I-55, in a backwoods of Yellow Bushy County, halfway to Memphis. It's pouring as we pull in, which thrills Mark. The kids spend every daylit moment here on four-track bikes, and the mud makes for wide, messy skids that splatter other riders.

The Hocutts, the Parkers, and a few other Parchmanite families put up money and bought a seventeen-acre hill here in the early 1990s, and Hocutt towed a bulldozer up and flattened the hill to create level acres for the two dozen or so trailer homes I see as we pull in. Hocutt points to his new double-wide and a big concrete pavilion he's put up beside it to shelter from the rain and heat. "You missed our housewarming party," he jokes as we park. "I coulda used you to get me up on bricks. There's my new antenna," he points to a small piece of metal on the roof. "That little piece of shit gets three hundred channels. Let's see how it holds up in this atmosphere."

After we settle in, Patty goes two houses down to visit Sarah Parker, the woman who cooked the fudge brownies for Jimmy Lee Gray's execution team. Donald, who hates hunting, gets on the couch with the remote to begin his weekend at Deer Camp. "I shot a fourteen-pointer, or some such, in 'ninety-one. But I'm not proud, believe me. It's like what Tom Mix said. 'It was the best shot I ever made, but it was the dirtiest thing I ever done.' Where's NASCAR on this thing?" I keep staring at his feet, which are white from the calf down. It makes him look like a Pinto.

"Donald, is that your gout?"

"That's my redneck tan. The white part's where my sock comes up to." He finds a heat in Michigan, where big names are competing for pole position in the Pepsi 400, then crosses his arms over his chest. In two minutes, his eyelids are drooping. Hocutt stays that way for much of the weekend. A dozen times, I could swear he's nodded off, but his attention is pricked. Four hours into the Michigan heat, the satellite signal gets lost to atmospheric pressure and Michael Andretti's car is frozen on the thirty-three-inch TV and Hocutt starts clicking. He stops at the History Channel, which is showing the multi-part documentary *The Nazis: A Lesson from History.* The segment on now is about the Einsatz Gruppe, the free-ranging death squads that from 1939 to 1941 shot hundreds of thousands of East European Jews, Gypsies, and Communists.

Hocutt tells me about getting flown last year to Cologne, Germany, where a local talk-show host, Hans Muser, had devoted a program to the death penalty. "They made a big thing out of quieting the audience down before I came out," he says. "I guess they didn't want to look like ugly Nazis or nothing, screaming for an executioner. You shoulda heard the cheers I got." Hocutt was on with an elderly German couple whose son had been beaten to death. "They were the tiniest little people, those old folks. They came back with me to my hotel, and just wouldn't let go of me.

They kept saying 'Danke, danke.' I didn't know how to say 'You're welcome,' so I just kept saying 'Danke, danke,' too. Ever since Jimmy Lee Gray, people have been wanting to touch me." His eyes open wide. "Look at that."

The segment we're watching doesn't pull any punches. There are twenty- and thirty-second clips of mass shootings, bodies falling into pits already filled with bodies, a photograph of a naked woman with an infant in her arms and a Luger raised behind her head. The scenes of mass graves get a full-throated "Shee-eesh" out of Hocutt, though he says he's seen the segment before. "You know how the condemned always say they're going someplace better when they make their final statements? I don't think they're going where they *think* they're going, but there has got to be close to a one-hundred-percent chance it's better than the shit they're leaving. Check this guy out," he cranks the volume. "This is what I call an executioner."

A thin, white-haired man from some former Soviet republic is being interviewed at the edge of a forest. He seems empathic and honest, and the smile on his face is constant and strangely warm as he discusses the mass murders during the occupation of his republic. Over his voice, the image cuts to sepia-tone stills of account-ledger pages covered in a precise Gothic script. Each page has a date on top and columns listing Slavic town and village names beside a number—87, 194, 274, 335, 1440.

"Those aren't death tolls, are they?" I ask.

"Um-huh," says Hocutt.

"And he saw it?"

"No, that's the sumbitch that did it. He killed his own people for the Einsatz Gruppe, then he turned snitch-jacket after the war."

The man, who spent twenty years in Siberia after testifying to a Soviet tribunal against townsmen who took part in the mass shootings, says he's paid for his crimes and has nothing to hide. But he speaks of the firing squads he manned only in the third person: *They* would work themselves up so *they* could do it. *They* would get more vodka than the others, so *they*'d be brave enough. "It's not easy to kill children," the man says.

The interviewer asks how the man felt while killing children, and is told, "I won't talk about that." The interviewer presses, and the man shakes his head with the strange, kindly smile. "You won't get me to say anything about that."

I tell Hocutt I've seen him smile that way many times—when he talks not about executions, but about the violence of running a prison, which he's often told me was far more "cruel and unusual" than any asphyxiation. I ask which of the two was more difficult for him and he smiles. "You wouldn't mistake either for a picnic."

"Which led more to your health breaking down eventually?"

He starts to answer several times, then gets lost in thought. His eyes close again and his jaw goes rigid. I think of Yeats: "Too long a sacrifice can make a stone of the heart."

"Donald, you look like you're two hundred years old."

"I see that in the mirror sometimes, too. I want to answer your question, but I'm thinking about everything, and I can't separate out what caused this and what that. When I went to Parchman, I didn't honestly believe there was anyone in the world that cared for me. I got extremely drunk, acted up, did some things I definitely shouldn't have, and never let anyone get close to me. I have my distance that I need to keep, and after years and years of distrust, it's hard to let it go. I know my wife and kids love me. My work is to let them. For eighteen years, my life was the Mississippi State Penitentiary at Parchman, then I got sick unto death. In the years I have left, I want the opportunity to do things with Mark that my dad never had a chance to do with me."

"You still feel you have a death sentence hanging over you?"

Hocutt's face goes blank and his eyes come down. "I'm thinking about a time I had a stunt plane go dead on me over Unit 27. There was no doubt in my mind for a few seconds I was going to die. I crash-landed, put the plane on its nose, and got kind of roughed up and groggy. But I didn't blink until I was on the ground, trying to get my bearings. I just looked down and saw the ground getting closer and closer. Death has been walking with me for a long time.

"I don't believe my work with the chamber caused any of my

conditions, but I remember feeling the particular kind of depression I suffer right after Edward Earl Johnson. That was the first time. And when I thought just now about that plane going down, I thought about Edward Earl sitting in the chamber. He didn't blink either. He's the only one of the four I ever think about. I mean, *ever*."

I'm stunned. Hocutt hasn't so much as pronounced the last names of any of his victims in a personal way since we've met. "Why Edward Earl Johnson in particular?" I ask.

"Maybe because the acid bath took so long to cook for him, and that was my job, and he had to sit there all that time. You're killing this guy, and you know that he's a dastardly fellow, but the worst feeling imaginable is if you're doing your job wrong. And maybe because we were talking to Mark before about the chamber, and Edward Earl was so small. It made me think about Mark being in there. That's why I don't believe in televising these things. You can't believe how hard it is on the families. Connie Ray Evans's mother died right after he did."

"What about Edward Earl Johnson's claim of innocence?"

The eyes glaze over, and the somnolence comes down again.

"Do you think he was innocent?"

"No, he did it. He led those sheriffs to the murder weapon." The eyes open wide again. He has a look I've never seen before: sadness mixed with alarm and an embarrassed smile. "I hope he did it."

"Donald, I know you became an executioner by happenstance, but in the end, is it something you *wanted* to do?"

He thinks long and hard about the word *wanted*. "That makes it sound like it was something I enjoyed doing. Those are the longest minutes I ever lived through."

"Would you do it again?"

"They'd have to pay me a lot more than five hundred dollars."

"Have you ever wished you hadn't done the ones you did?"

"Never." He shakes his head. "I regret many things I've done in my life, but I have not lost one ounce of regret to the men I killed."

•   •   •

Edward Earl Johnson's possible innocence, I tell Don Cabana, is very much on my mind as I sit at the dining-room table of Cabana's ranch house in the university district of Hattiesburg. I've driven eight hours from Deer Camp in relentless rain, and there's no table I'd rather be sitting down to. Cabana, who gives me a glass of sweet tea and excuses himself to take a long-distance call, is an intensely sympathetic man. After five years on the road, teaching, debating, or testifying for the abolitionist cause, he's also a professional interviewee; over the years, he's seldom failed to give me exact, forceful answers, even to questions that are inexact or reaching.

I have a particularly unformed question this afternoon, which, for reasons I don't yet understand, has been plaguing me, off and on, during my white-knuckle drive down to Hattiesburg. It's actually a few lines from Sophocles, and they come whenever I picture Hocutt talking about Edward Earl Johnson sitting and waiting in Parchman's chamber: "Who is the victim? Who is the slayer? Speak!"

"Funny you should mention Edward Earl," Cabana says with gentle irony when he gets off the phone and sits across the table. "I came back from teaching not long ago, and my son told me Bill Allain, the old governor, was on the phone. He was apparently having some strong misgivings about Edward Earl Johnson. He asked if he could drive by and we talked about it. All these years later."

"What did you talk about?"

"Maybe someone had come to him with information—I'd rather not go into it—and maybe he'd finally met enough of our local sheriffs to know some were capable of what Edward Earl claimed was done to him in the woods while they were driving him to Jackson."

Cabana begs off when I press for details, and I settle for asking why it's so difficult, so many years later, for men like Hocutt and himself to discuss these executions.

"Being part of an execution team," he explains, "is the ultimate male bonding. Breaking those strong, silent ranks is something you're strongly conditioned not to do, even when you're as opposed to the penalty as I became, or have gotten as alienated from the

DOC as Hocutt has. I broke rank because I had a bond with God that was stronger. And if there was any doubt in my mind, my body let me know."

Cabana has had three massive coronaries in the last five years. The first was in November 1993. He recalls it impressionistically: the starched sheets of the hospital gurney and the ugly brightness of fluorescent lights above him just before he went under the knife, the sound of his voice as he screamed that he had to see a priest. For five years, Cabana had been sure that he had accepted Connie Ray Evans's eleventh-hour absolution from the black seat of the chamber. Facing death, he says, he no longer had that certainty. Neither did the priest, who, as luck would have it, was both painfully honest and an abolitionist. The priest told him he really didn't know what to say.

"Give me an answer to my question," Cabana yelled. "Please. I have to know."

"I truly cannot answer," said the priest. "I don't think you have to worry about God holding you responsible, but I don't know. God considers all life sacred. Perhaps the answer is simply what was in your heart at that moment."

I tell Cabana I've been looking for the real motive behind the death penalty for years, yet somehow that simple question never occurred to me. He smiles. "Wouldn't that be a little hackneyed?".

"Let's see. What did you feel when Connie Ray Evans died?"

"I felt compassion and mercy at the moment I asphyxiated him. In fact, I felt love. I don't know if that makes me an innocent man. But I'll take it."

"I asked Hocutt this: Do you feel like you're living under a death sentence?"

"Three coronaries," he says. "Quadruple bypass the last time. I know a thing or two about death sentences."

"Do you connect that to the death sentences you carried out?"

"Nah," he says. "I'll break rank and tell tales out of school about the gas chamber, but I'll be damned if I'm going to send the State of Mississippi a moral bill for my heart condition. Besides, I do have a real conviction"—he grimaces halfway through what he now realizes is a pun—"I'm an innocent man. There's no guilt breaking my heart."

"What about Edward Earl Johnson?" I ask.

"What about him?"

"What did you feel about his innocence?"

"I didn't think for a moment about his innocence or guilt at the moment I killed him."

"And now?"

"And now, I try not to think about it, but when I do I pray he was guilty. I pray he did shoot that policeman, in the coldest blood imaginable. I went through his file over and over for something to give the governor. And believe me, no one would've been happier if I'd found something there than Bill Allain."

"And what did you feel at the moment of his death?"

"At the moment Edward Earl Johnson died." Cabana tightens his arms across his chest and thinks hard. "As his jailer, as the man who killed Edward Earl Johnson, I felt bad I couldn't stop him from soiling his pants. He was hard to get to know, but I knew that getting out of the red death row outfit and into jeans had meant a lot to him."

"That's a little impersonal, isn't it?"

"Perhaps. But Edward Earl *was* hard to get to know. Real hard. Something about him kept you at a distance."

"Kind of like Hocutt?"

"How do you mean?"

"Hocutt told me: 'I have my distance that I need to keep.' I get the feeling Hocutt's pretty haunted by Edward Earl."

"I think so. Maybe some of the others too. Donald was always a suspect member of the goon squad. He ran Edward Earl's and Connie Ray's executions for me, but he was suspect even there. I don't know what goes on in that man's heart, but I know that it's warlike and that it is hell on him. For four or five years now, most of the people we have in common thought he'd be eating a gun any day."

I tell Cabana about the lines from Sophocles that have been going through my mind. He nods, then talks about testifying against capital punishment a few years ago before a joint committee of the Massachusetts Legislature, which was weighing a return of the death penalty to the commonwealth. "I gave them a modest proposal, should they decide on it. Have three executioners: the sentencing judge, the DA, and the jury foreman. And, like I said before, make the other eleven jurors be the witnesses. That way, you wouldn't make corrections people violate their profession by taking life, and you wouldn't have to deal with the press. It's embarrassing as hell to be in front of these people with blood on your hands. That may also explain why it's sometimes difficult to discuss it."

I ask if it also helps explain Mississippi's reluctance to execute post-*Gregg:* if the very publicness of the protocol has made it hard to carry out executions. "Perhaps. But you also have to factor in a liberal state supreme court. A conscienceful DA, much of the time. A circuit court that couldn't help but be concerned by the brutality of the chamber. The list goes on. Take them all away, and what do you have now? A governor, Kirk Fordyce, who finally got the chamber out and is doing everything he can to 'close the loopholes,' as he puts it.

"I really don't know what the motive behind the penalty is, or who the *real* executioner is you seem to be looking for. But I teach criminal justice for a living, and I know that the penalty comes in watersheds and thresholds that politicians build and open, just on or about election day. You can call them your real executioners and get no argument from me, particularly in times like this, when everyone from the county courthouse to the White House wants to milk the law-and-order vote. And once you've done that, you can say: It's the people that elect them; that the *yous* and *mes* that pull the ballot box levers are the hands that pull the switch. But it's not historically accurate. Democracies just don't execute as much as autocracies. The Athenians of Pericles' time executed far less than their neighbors and predecessors, western bourgeois democracies have executed progressively fewer people, and dictatorships and monarchies still have the highest rates of execution. They lop heads in public every week in Saudi Arabia, and there are few years that the Chinese don't shoot more people than we've executed as a nation since *Gregg.*

"And finally, the 'vox populi' fails for the same reason those insidious pro-penalty polls fail. Seventy-five percent say they're for it. For what? An eye for an eye? That's like two plus two equals four. Of course you're in favor. But offer them murderers getting life without parole instead of execution? The figure's down to the low sixty percents. Throw in a better understanding of mitigating factors, or having murderers work for the state or for the restitution of the victim's families, it goes well below fifty percent. Then you're back in the 'sixties, when people did think about the public weal and things that we now call 'mitigating factors.'"

"Wilbert Rideau told me he sat out the whole decade on death row, but he calls it a period of enlightenment, simply because the executioner went unemployed for the first time."

"The 'sixties didn't bring me to enlightenment," Cabana shrugs. "They brought me to Vietnam. Then corrections work, and eventually to being an executioner myself. You say you're searching for a real motive behind the death penalty, but you're probably only asking that because you've never seen war. There's nothing unusual about human beings killing each other. And put a capital E in front of Enlightenment, and you'll see great, humane men, John Stuart Mill, Jean-Jacques Rousseau, who were *eloquent* about the need to kill their fellow man. The folks that brought the Age of Enlightenment also brought the Reign of Terror and its aftermath. And I feel the same duality myself: I go around the country as an abolitionist, sweet voice of reason and tolerance and all, but don't forget: I didn't just kill once, or even twice. I slapped the teetotaling shit out of a number of convicts. Not that I'm proud of it or anything."

"You don't sound proud. You sound guilty. Before, you said you felt like an innocent man."

"Maybe because you got me thinking about Edward Earl again, the night he was asphyxiated. I was in the shower two hours later, scrubbing and scrubbing. Then I showered again. I just couldn't get the sweat and grime off me to the point where I felt clean enough to go to sleep."

Cabana's arms tighten on his chest and he starts rocking. "It hasn't come off me yet."

Heading north to Memphis the next morning, I stop at a favorite barbecue pit outside Jackson. The storm is over, the sun is shining in a perfect blue sky, and I've mercifully stopped thinking about gas chambers, guilt, and Sophocles.

Halfway through my Smokey Joe, however, I find myself in yet another reverie about death and murder. It starts with a memory of an interview I did six years ago in the office of a Russian police investigator, Andrei Burakov, who after a decade-long pursuit had trapped and arrested Andrei Chikatilo, the serial killer who raped, murdered, and butchered the genitals and inner organs of fifty-four children in the woods outside Rostov. Burakov had been throwing crime-scene photographs across his desk for me—photos that were hard to look at but also hard to turn away from. Chikatilo not only killed savagely but left bodies in a way that graphically *exposed* the mutilated organs, as though they were being presented for metaphysical exegesis. Burakov had started talking about a nervous breakdown he'd suffered halfway through the pursuit. "I would get up in the night, maybe awake and maybe still sleeping, and walk through my house. Then I'd come to my son's room and look in." He stopped talking and gazed at the trees outside his office window. "By day, I go to the woods and find the children he left," he said. "At night, I leave my bed and kill them myself."

There have been four moments, while researching this book, when my goal of understanding the true motive of the death penalty seemed at hand. They came when I'd most clearly apprehended the coups de grâce of the four asphyxiations I've chronicled, and saw the executions through Hocutt's or Cabana's eyes. The understanding wasn't far from what I had gleaned from the abjection on death row, or what I had learned in the *Angolite*'s offices: We execute to exert power over what horrifies us most supremely, and we execute imperfectly—randomly, cruelly, unusually—because murder itself seems exactly so to civilized eyes. In a state of nature, there is nothing random, cruel, or unusual about killing.

At such moments, however, I was able to look at these asphyxiations without judgment, and they seemed anything but uncivilized. In fact, they seemed like an extreme form of "social" behavior—as precise as the Gothic script listing the Einsatz Gruppe's death tolls on an accountant's ledger, and as formal as Parchman's execution-team members calling each other *Mister* as Jimmy Lee Gray's final hour approached. In such epiphanies, there was very little about killing, murder, or an execution that was alien to me. All were explicable by the same syllogism: Killing is part of nature; murder is a human act; an execution is in the deepest weave of the social contract. Not an "expulsion of taboo," an execution seemed to be an embrace of taboo, under the control of a state protocol.

While it is indeed "a terrible business, to mark out a man for the vengeance of man," it is also a very strange business to be able to look at taboo directly and without judgment. It instills a feeling of almost divine wisdom, which in former times was a succinct part of an execution's covenant. In execution rites of the Middle Ages, the condemned were expected not only to forgive the executioner but at times to physically embrace him. Execution was punishment but also a kind of marriage: two humans joined under a bond that both understood and transcended the actions they were about to take. The vows of love that passed from the chamber door between Don Cabana and Connie Ray Evans are a modern example.

This quasi-omniscience fails, however, when I think of Donald Hocutt considering Edward Earl Johnson's possible innocence. As the saying goes: It puts all heaven in an uproar. Suddenly, murder and executions no longer seem natural. They seem like abominations; inexplicable even to the men who investigate them or carry them out. They make policemen in Rostov and executioners in the Mississippi delta walk through their houses in the middle of the night, unable to sleep or wake, looking at their small children they're convinced they'll be murdering in the morning, or searching for something even smaller in their kitchen cabinets.

NOTES

PROLOGUE

the Midnight Special . . . The ballad of the same name originated with the legend among Parchman inmates that the first man to see the "ever-loving light" of the midnight train would be the next to ride it, out on pardon or parole.

INTRODUCTION

Why do we want *to do it?* . . . Of the 66 percent of Americans in favor of the penalty, conducted by the Harris Poll on February 14–15, 2000, 46 percent said they felt it was the only punishment that fits the crime of murder; 12 percent favored it because it "saves the taxpayer money"; 8 percent cited deterrence; 5 percent said the condemned "deserves" it; 4 percent said executions stop the condemned from repeating the crime; 3 percent cited the Bible.

•   •   •

"a badly orchestrated opera" . . . Robert Weisberg, quoted in *Crime and Punishment in American History*, by Lawrence M. Friedman (New York: Basic Books, 1993), p. 316.

ordered a videotape made of the 1992 asphyxiation of Robert Alton Harris . . . Probably the only visual record of an asphyxiation, the tape was sought after for two years by collectors, news organizations, and filmmakers, notably one documentarian who sued the State of California for the right to film executions and market them for pay-per-view. Judge Patel ordered the video kept in a safe in her clerk's office, allowed no copies made, then ordered its destruction in 1994.

which would remain the most common form of American execution . . . Seventy-four of the first hundred post-*Gregg* were by electrocution.

Prior to New York's conversion to electrocution in 1890, the state commissioned a history of execution methods, included in a more general *Report of the Commission to Investigate and Report the Most Humane and Practical Method of Carrying into Effect the Sentence of Death in Capital Cases*. The author, who clearly warmed to his subject, delineated forty-five basic forms—not counting variations, which would have taken him into triple digits. He moved from elemental methods (stoning, disemboweling, burning, boiling, forced ingestion of boiling liquids, strangling, drowning, cutting, burying alive, and crucifying) to the medieval (bleedings performed in spiked boxes and barrels of many varieties, the French torture-murder known as the *peine forte et dure*, in which the condemned was covered with heavy stones and crushed over a period of days) before narrowing to then-contemporary methods. He chose the four he considered most humane: electricity, guillotine, the garrote, and poisoning by a syringe of prussic acid, which in practice branched off to the twentieth century's offerings, the gas chamber and lethal injection.

Consciousness is a relative thing . . . The humanity of the guillotine, likewise meant as a humane alternative to the ax, went unquestioned for a century after its use in the Reign of Terror. (The head-chopping device, incidentally, did not begin with the French Revolution: the first mechanical blade was of Irish origin, from the fourteenth century.) In 1880, a Dr. Dassy de Lignières received permission to attach the condemned's severed head by transfusion to the veins of a living dog. He allowed the blood to flow for several minutes, and came to the conclusion that the brain was sensate long after execution. "The head," he wrote, "separated from the body, hears the voices of the crowd [and] feels himself dying in the basket." An early twentieth-century doctor named Beaurieux revived an executed man simply by calling his name, "in a strong, sharp voice: Languille. I then saw the eyelids slowly lift up—such as happens with people awakened or torn from their thoughts. Next, Languille's eyes very definitely fixed themselves on mine and the pupils focused themselves. I was dealing with undeniably living eyes, which were looking at me." In 1956, the French Academy of Medicine condemned the method: "Death is not immediate. Each vital element survives decapitation to some extent. Death by guillotine is a murderous vivisection." The academy added that this murder was in turn "followed by a premature burial." An assistant executioner once told Albert Camus that the hands of headless men struggled against the cords binding them well after decapitation. The body of one man, he said, was "still shuddering" in the graveyard twenty minutes after the sentence.

chemical entombment . . . There are several cases of clearly painful lethal injection. Stephen McCoy, executed in Texas, May 1989, reacted so violently to his injection one of the witnesses fainted, knocking over two other witnesses. Robyn Lee Parks began to convulse two minutes after his injection, in Oklahoma, March 1991. Muscles in his jaw, neck, and abdomen went into spasm and he gasped and gagged for more than a minute.

The possibility of undetectable pain in the electric chair was raised by experts from George Westinghouse to Nikola Tesla. Dr. Harold Hillman, an English physiologist specializing in electrical burns, provided the most detailed view of this possibility during testimony before a 1990 investigation by the Fifth Circuit Court of Appeals into the constitutionality of Louisiana's electric chair. "Although it is widely believed that the person being electrocuted loses consciousness and the sensation of pain immediately, there is no scientific evidence whatsoever to support this belief. The massive electric current stimulates all the muscles to full contraction. Thus the prisoner cannot react by any further movement, even when the current is being turned off for a short period and the heart is still beating, as has been documented in numerous cases. . . . It is usually thought that the failure of the convict to move is a sign that he cannot feel. He cannot move because all his muscles are contracting maximally."

CHAPTER ONE

Sixty percent were carried out within a two-hour drive of Interstate 10 . . . The statistic jumps to 70 percent if one adds the neighboring states of Georgia and Arkansas, the state prisons of which are both within a day's drive of I-10.

"everyone gotta dance/with the Grim Reaper" . . . Hip-hop has become the lingua franca of death row. Robert Alton Harris, who quoted these words prior to his 1992 San Quentin asphyxiation, had picked them up from the movie *Bill and Ted's Bogus Adventure,* which he'd seen in prison.

escaped death by becoming the hangman . . . One early English statute required that the hangman be a criminal sentenced to death and granted reprieve only so long as he performed as executioner.

dark, obscure, the locus of a pariah . . . Provincial European executioners often lived out of town, in isolated though not necessarily inferior circumstances. In church, they and their families were generally assigned a separate pew; if they ate in the tavern, they were seated separately. In cities, this distance was more marked. The seventeenth-century Nuremberg executioner Master Franz

Schmidt, known to posterity for the laundry list of a journal that he kept of his executions, was given a large stone tower on a spit on the Furth River. The property came with a gated portion of one of the two bridges connecting the spit to the city, which was his sole walk-way for the early-evening constitutional. If the executioner had been drawn from the criminal ranks, the man was physically "marked." In the Swedish town of Arboga, in 1470, a thief on the gallows, commuted for serving as hangman for his fellows, was branded with an iron. Two centuries later, a thief in the town of Gronso agreed to become his fellows' hangman; both his ears were cut off.

he took after his mentor Edwin F. Davis . . . Until his retirement in 1914, Davis appeared in a Prince Albert coat and black felt hat at all his executions. Once asked by a condemned man to come to his cell, Davis agreed. "Well, how do I strike you?" he asked.

"You're the ugliest son of a bitch I've ever seen," the man said.

While Davis was adjusting the head electrode in the chamber half an hour later, the man turned to him and said, "I still think you're the ugliest son of a bitch I've ever seen."

the executioner's name is public record only in Mississippi . . . and Alabama . . . San Quentin's warden, Dan Vasquez, pulled the switch on both California asphyxiations, but this fact was made public only in the course of an ACLU lawsuit. Prior to abandoning its electric chair, the State of Indiana had the warden pull the switch.

One of the last states to use a civilian executioner . . . also one of the last that hoods the man . . . Lethal injections in Pennsylvania are done by civilian paramedics, though a Harrisburg physician is officially listed as the state executioner. Oklahoma hoods its three lethal injectioners.

An anonymous man referred to by the *New York Times* in 1990 as "the last civilian American executioner" is Delaware's former hangman. The story labeled him a "backwoodsman in Canada," reachable only by a message left on a certain tree stump. In what may have been a lapse of security, this hangman was named—as a "Mr. Ellis"—in a 1991 response by the State of Delaware to a motion filed for the condemned murderer Billie Bailey. "Ellis" may have been a pseudonymous borrowing from Canada's twentieth-century hangman, Arthur Ellis, itself a pseudonym for Arthur Bartholomew English. Pseudonyms for hangmen were common. For centuries, all English hangmen were named Jack Ketch, the first national hangman, though Ketch was a notorious bungler. Delaware eventually told the court that DOC employees would be trained for the job.

In 1989, spokesmen for the State of Washington, the first to hang post-*Gregg*, said they had contracted "one of two hangmen" still living in the United States, for $1,500. Their choice between the two, they later said, was made easy when one suffered a nervous breakdown. The condemned won a stay, and the contract was voided. It would probably have been the highest fee ever paid for an execution. An independent film producer who contracted with the condemned Westley Allan Dodd to film his 1992 hanging waged a lawsuit to have the hangman's identity made known to him, but nothing came of it except an extreme tightening of information about the protocol. Washington spokesmen have since been told to say nothing about their hangman, except that he is issued a contract with each hanging, as per a territorial statute still on the books.

Walla Walla hangings are carried out by the *US Army Manual*, which recommends "six coils in a knot fashioned at one end of 30 feet of manila hemp, boiled beforehand to take out stretch and any tendency to coil." They take place behind a window in a room on the second flight of a two-story indoor gallows. The condemned, dressed in new prison-issue jeans, a T-shirt, and sneakers, is walked forty feet from one of the last-night holding cells to one of two nooses, where he appears at the window and is offered a chance to make a statement to the witnesses, assembled in a room on the ground floor. Levolor blinds are then drawn over the window, and two men appear, visible only as silhouettes. One is the hangman, who is seen placing a hood around the head of the condemned. In Washington, the condemned is strapped to a board

to guarantee sufficient weight to cause death by a quick disloca-
tion of the vertebral column, commonly known as "snapping the
neck." Then he adjusts the noose under the left ear (under the
right ear can cause prolonged strangulation) and pushes a red button that springs the trap. One Walla Walla official tells me he's
forbidden even to say what type of mechanism springs the trap,
but adds, unbidden, that, should a double execution ever occur,
there are two iron eyebolts suspended from the ceiling in the gallows room.

Charles Rodman Campbell, hanged in 1994, was dragged the
entire forty feet from the last night cell to the gallows.
Campbell, who had vowed, "If I go, correctional personnel will
go with me," was lying face-down on the floor of his cell when
the hangman came for him. After refusing to get up and come to
the cuff port to surrender his wrists, he was pepper-sprayed by
guards, then bound and cuffed, toweled off, and dragged to the
gallows.

whose identities are also secret . . . Though the Florida protocol—
from the voltages used to the tick-by-tick timetable leading up to
and following the 7 A.M. electrocutions—is publicly documented,
the veiling of the larger execution team is extreme. They may change
the team periodically: following the 1996 execution of José
Medina, whose head caught on fire, a "corrections committee"
ordered an investigation, published as "Report on findings and
Recommendations Prepared following visit to Florida State
Penitentiary at Starke, FL on April 8, 1997." "It is noted," a visiting specialist wrote, "that there was a significant change of personnel between my visit in 1990 [following the botched execution of
Jesse Tafero] and 1997." The report, commissioned by the state
during one of its intermittent studies of switching to lethal injection,
contains the basis of and in many cases extended quotes and minute
details from the protocols of the thirty-eight states that were then
executing. It is available on the Web: www.dos.state.fl.us/fgils/agen-
cies/fcc/reports/methods/.

He sits there until sunrise, when he's summoned . . . The Greek
national executioner, now retired (a capital punishment statute is
on the books but hasn't been used in decades), worked in similar
anonymity. His name was known only to a handful of officials, and
he lived on a "secluded island," its name also secret. Once each
year, he took an overnight ferry to Piraeus, arriving at 7 A.M. of
"national execution day."

*the plate glass window behind which America's most prolific exe-
cutioner stands . . .* A comprehensive and otherwise reliable
account of the first lethal injection, of Charlie Brooks in
Huntsville in December 1982 (*Official Detective*, August 1983),
said the executioner was "a prisoner medical technician, otherwise unidentified." The famously unreliable execution technician
Fred Leuchter, self-styled engineer and Holocaust denier, who
designed post-*Gregg* execution machinery for a half dozen states,
said, in Stephen Trombley's *The Execution Protocol* (New York:
Crown, 1992), that Huntsville's warden served as Brooks's executioner. The use of inmate labor in any post-*Gregg* execution is hard
to credit, though inmates were employed in various aspects of executions before *Furman*. Huntsville's warden could not have performed the execution as he was in the death chamber at the time
Brooks was injected.

"none knows the day or the hour" . . . The French appear to have
been unique in keeping the condemned from knowing the exact
date of execution. Very early on the morning of execution, he was
gently roused by two guards, who did not wear shoes and went in
for him at the last minute. One sees this quite clearly in gruesome
footage, shot clandestinely from an atelier window, of the last
guillotining in France, included in the A&E documentary *The
Executioners*. The condemned, who is quickly led out a doorway
and up three stairs, is rumpled from a night's sleep and only grad-
ually conscious of his impending execution, which is conducted
with extreme haste.

the Hippocratic Oath ... The AMA's directive applied directly only to lethal injection, though physicians are employed equally to determine death in the other four forms of execution. The American Psychiatric Association issued a similar directive in the case of thirty-six-year-old schizophrenic Michael Owen Perry, a Louisianan who murdered his parents, two cousins, and a two-year-old nephew whose precocious intelligence convinced Perry he was the devil. The APA's directive, occasioned by the state's use of the antipsychotic Haldol to maintain Perry's "competence" to face execution, was far more inclusive than the AMA's: "The physician's serving the state as executioner, either directly or indirectly, is a perversion of medical ethics and of his or her role as healer and comforter." On September 6, 1991, Donald "Pee Wee" Gaskins cut his wrists several hours before his scheduled electrocution in South Carolina's Broad River facility. The resident physician sewed the wrist with twenty stitches, and Gaskins was executed on schedule.

Departments of corrections go out of their way to appropriate medical language, equipment, and imagery: The phrases "standard hospital issue," "execution-technician," and "minimally invasive" occur repeatedly in execution manuals and guidelines, as does the word *generic*: "the generic drugs based on University of Texas School of Pharmacy research to find the most effective and quick-acting agents" (from Arkansas's protocol); the "generic hospital gurney" with "three-and-one-half-inch pad with extended arm" (from Oklahoma's); even "generic leather restraints" (from the above-noted study by the Florida DOC for potential lethal-injection protocols).

lost temporary access to doctors ... American physicians and scientists have had a long and intimate relationship with executions. When the United States Army hanged thirty-eight Sioux warriors after an 1862 Minnesota uprising, the mass burial site was looted by local doctors and scientists for vivisection; one was William Mayo, father of the two men who later founded the Mayo Clinic. Fourteen of the twenty-five witnesses to the first electrocution, of William Kemmler, were doctors. The sight of them so unnerved Kemmler he begged his jailer, Deputy Sheriff J. C. Veiling: "Don't let them experiment on me more than they ought to." The prevailing theory of death by electrocution at the time—the "suspended animation" theory—was that the jolt "suspended" life, and that it was the autopsy itself that caused death. In essence, those who believed the theory came to witness the autopsy of a live person. *the free-world attendant was a justice of the peace* . . . The IV team in Arkansas is made up entirely of volunteer local medical personnel, though they are supervised by "a DOC Medical Administrator who purchases IV lines, drugs, other medical materials and packages and keeps them under lock and key." During executions, this administrator stands beside the condemned, wearing a low-frequency headset to communicate with the executioner.

offers no sedatives . . . Virginia not only offers but requires a sedative: a ten-milligram intramuscular injection of Thorazine, one hour prior to execution. The Nevada "Execution Manual" also mandates "pre-medication" an hour before execution, to prevent last-minute resistance. Because Pavulon can lead to spasmodic coughing, some states medicate the condemned with a simple antihistamine an hour before sentence.

The warden at Starke in Florida shared a tumbler of whiskey with John Spenkelink in his cell on the morning of electrocution, but the practice was soon tabled. It marked the official end of perhaps the longest tradition in the history of executions: getting drunk. Executions of Greeks and Romans of property were often preceded by multiday feasts. The underground chambers in the seventeenth-century prison at Nuremberg, an early form of death row, was the scene of all-night parties lasting until moments before the morning execution. In London, the road to the hanging-tree at Tyburn passed a series of pubs where the condemned was treated to drinks, though the practice ended with the removal of hangings to prisons—largely to cut down on the drunken rioting that often occurred.

As late as turn-of-the-century America, prisoners were at times allowed to drink themselves silly before hangings. Frank Brenish, a wife killer in Tennessee, went so far as to protest in a drunken slur from the gallows: "They oughtn't to hang a man when he ain't to his right mind!" He was hanged regardless, though the sheriff did honor Brenish's other request, which was to hang separately from three black men he was originally scheduled to swing with.

Rumors have persisted: graphic tales of a drug overdose/suicide attempt and a last-night drinking party by Gary Gilmore and his girlfriend made the rounds among wardens at the American Correctional Association convention following the execution. Blood and urine samples from the autopsies of men executed post-*Gregg* in Florida have shown significant amounts of amphetamines, alcohol, and marijuana.

CHAPTER TWO

when he finally does send him to death . . . Antonio James was executed on March 1, 1996.

thirteen loops on a hangman's knot . . . Nooses, Woody Guthrie's song notwithstanding, do not usually have thirteen loops, nor do scaffolds necessarily have thirteen steps. The morbid tradition may have originated in nineteenth-century England, where hangman William Marwood used a thirteen-foot length of Italian silk-hemp rope bound with chamois leather. The use of the number thirteen was reinforced in American lore by the scaffold in the Western District of Arkansas, to which the "Hangin' Judge," Isaac Charles Parker, sent some eighty-eight men. Parker's executioner, George Maledon, the "Prince of Hangmen," used thirteen loops on his knots, though there were only twelve steps up the gallows. Before the electric chair was introduced to Huntsville, the condemned were hanged in the basement below death row, which was reached by a staircase of thirteen steps. The same was true in San Quentin.

"what it sanctifies is denial" . . . The brothers of Santa Maria della Croce al Tempio, a medieval fraternal order that volunteered their services to men on the gallows, held a large, double-paneled painting, called a *tavoletta*, in front of the condemned's eyes as he

mounted the steps to prevent him from seeing the executioner and his instruments.

CHAPTER THREE

pale blue in gaseous form . . . The German word for prussic acid is *Blausäure;* the Greek root, *kyanos,* means dark blue. The cheeks and necks of hanged men tend toward this dark blue, rather than the purple normally found in violent death, and are said to become "cyanosed."

crushed or bitter almonds . . . A condemned man dying in San Quentin's chamber in the 1940s called out to Warden Clinton T. Duffy that the cyanide smelled "just like rotten eggs." That was the sulfuric acid.

waves of unnatural flatness and freakish-looking tightening . . . Similar rippling movements of the body and limbs are seen in hangings. I have heard that the phrase "well hung" comes from the fact that hanged men taken from the gallows had erections, but I have had no luck confirming this, either physiologically or etymologically. It is documented, however, that traces of semen are found in the underwear of the condemned after death, along with urine, blood, and feces. The Sadean connection between asphyxia and heightened orgasm, which fills the literature of the fetish, was reiterated in a series of apparent suicides by hanging among teenage boys devoted to heavy-metal music in the 1980s. The "suicides" were later discovered to have been accidental—incurred by the boys attempting to achieve extreme orgasm at the end of a rope.

the oxygen-starved heart and brain finally fail . . . For a more detailed analysis of the physiological and chemical effects of asphyxiation see http://www.idiom.com/~drjohn/cyanide.html.

Bedbugs, mice, rats, cats . . . rabbits . . . piglets . . . California's chamber was first tested on a twenty-five-pound piglet, two weeks after it had been brought across the bay on a barge in December 1938. "A pig's the hardest thing to kill," said the test's designer, Horace Jackson, an engineer who may or may not have designed the chamber for Eaton Metal Products, then based in Denver. "If it works on a pig, it'll work on a man." They laid the animal across the chamber's two stainless-steel chairs, sealed the chamber, and dropped the lever. Newsmen wrote that the pig seethed for three minutes before dying in a "drooling paroxysm." Willis O'Brien, the San Francisco *Chronicle* reporter on the scene, called it "the most hellish form of capital punishment since civilized courts sentenced one to be hanged, drawn, and quartered." O'Brien, who deserves high marks for knowing that drawing and quartering was typically preceded by a partial hanging, which left the victim half-conscious, went on to seethe himself: "If the mercy of Nepenthe comes as slowly for the human body as it did for the little porker, then there will be terrible things done to men's souls and their tortured brains."

The first animal to die with a measured shock of electricity was a seventy-six-pound Newfoundland, brought to the lecture hall of the Columbia School of Mines in 1888 by a man named Harold Brown, an early PR flack/electrical engineer who aligned himself with Thomas Edison in a monthslong battle to have the alternating current of George Westinghouse used for the electric chair, rather than Edison's direct current. Electricity was a terrifying prospect in its infancy, and proving a hard sell. Both Edison and Westinghouse wanted the other man's current used for the electric chair, and thus identified in the public's mind with unsafeness. The dog Harold Brown brought to the lecture hall suffered terrible burns and pain after being subjected to a blast of Edison's direct current but did not die, which was Brown's intent. When the audience demanded the dog be put out of his misery, Brown used Westinghouse's alternating current.

Animals have been used as executioners, have been executed themselves on gallows, gibbets, chopping blocks, and electric chairs, and have been the cause of death sentences for acts of bestiality and theft. A mare sodomized by a member of the Mormon Militia in

Echo Canyon, Utah, in 1857, was put to death, for example, though her sodomizer, condemned to death at the same trial, was merely exiled. At times, animals who have caused human deaths have been tried and executed, dressed in men's clothes for the ceremony. If the theft of an animal was grounds for execution, the stolen animal has been used as the unwitting executioner. In seventeenth-century Scotland, the blades of both the Halifax Gibbet and the Scottish Maiden were tied by long ropes to the poached beasts and tripped when the animal wandered off. The emperor Nero, who fattened his prize eels on slaves and fed condemned men to such animals as lions and rats, had a favorite "animal" executioner, presented to him by an artisan hoping to curry favor. It was a hollow metal sculpture of a bull with a brazier for a stomach, big enough to fit a man. Reeds and pipes in the bull's neck made the screams of the man roasting inside the stomach sound like a bull's braying. A practice in precolonial India was to lay the condemned's head on a wide, flat stone, where it was crushed by the foot of a trained elephant. The crowd-pleasing hanging, drawing, and quartering was a spectacle that accompanied bear baiting in France, where that form of execution seems to have originated. The practice of having the man quartered by four horses, who were each attached to a limb of the condemned and whipped to ride off in different directions, was ended when objections were raised by noblewomen concerned by the strain put on the horses. The famous Moroccan form of execution—burying the condemned in ant-ridden sand with his head smeared with honey—persisted into the twentieth century.

seems, to the naked eye, to go on for minutes . . . Aron Mitchell, the last man asphyxiated pre-*Furman* in California, took over twenty minutes to die. He also had a recorded time of over five minutes before losing consciousness.

have all been survived and described . . . An Iranian, identified by the daily newspaper *Kayhan* as Niazali, survived hanging in February 1996. "That first second lasted like a thousand years," said Niazali.

"I felt my arms and legs jerking out of control. Up on the gallows in the dark, I was trying to fill my lungs with air, but they were crumpled up like plastic bags." *Kayhan* reported that Niazali struggled for twenty minutes before he was cut down. A John Smith, hanged for fifteen minutes at Tyburn on Christmas Eve, 1705, told his rescuers: "When I was turned off [from a cart] I was, for some time, sensible of very great pain occasioned by the weight of my body and felt my spirits in strange commotion, violently pressing upwards. Having forced their way to my head I saw a great blaze or glaring light that seemed to go out of my eyes in a flash and then I lost all sense of pain. After I was cut down, I began to come to myself and the blood and spirits forcing themselves into their former channels put me by a prickling or shooting into such intolerable pain that I could have wished those hanged who had cut me down." Seventeen-year-old Willie Francis of St. Martinsville, Louisiana, survived his 1946 electrocution in the state's traveling electric chair, then sat in prison for a year while appellate courts and the supreme court decided that the state had the right to execute him again, though sentence had been carried out: Francis had been hit with the full 2,000 volts required by his death sentence. Francis said green, pink, and yellow speckles came to his eyes when the current surged through his body and that his mouth tasted like peanut butter.

to describe the experience of being gassed in the chamber . . . Long after the gas had risen on the first two men to die in San Quentin, one seemed to mouth the words, "It's terrible." The resident priest, who grew to hate the method, said he heard the other man say, "It's bad." The third man to die in San Quentin (he went eight days later) seemed, to a *Chronicle* reporter, to be mouthing the words "Too slow." Caryl Chessman, before entering San Quentin's chamber twenty-two years later, told reporters he would nod if the gas was painful. (Witnesses in San Quentin saw only the back of the condemned's head.) Beginning at the sixth minute, when he seems to have reawakened, Chessman's head nodded for several minutes.

In a double execution in Nevada's two-seated chamber, two Hispanic prisoners talked heatedly as the gas rose, but none of the witnesses understood what they were saying. James Stephens, a Navajo warrior executed in Nevada in 1941, let out a war whoop as the gas rose. Holding his breath for over a minute, he was able to free one hand, then remove first the blindfold that had been forced on him and then the other straps. Turning to the witnesses, he stared at them for a few seconds, leaned back calmly in the chair, and breathed deeply. In 1949, Leanderess Riley, a thirty-three-year-old, suffered probably San Quentin's most extended and wrenching asphyxiation. He was a deaf-mute. Weighing only eighty-eight pounds, he had been dragged the entire last mile, screaming unintelligible syllables. Though he was strapped in with the tightest cinch, his wrists and ankles were too thin to be contained. Twice able to free himself before the cyanide pellets were dropped, he ran from window to window, screaming at the witnesses, until the door was opened and he was strapped in again. When he broke free a third time, the warden told the executioner to proceed. Riley screamed and ran again as the gas came up, then finally died in front of one of the windows. Carl Austin Hall and Bonnie Brown Heady, kidnapper/murderers, were strapped down in Missouri's two-seated chamber in 1953, and Hall broke down. Heady, who had been screaming obscenities at her executioners throughout the strap-down, stopped long enough to tell Hall, "Shut the fuck up and take it like a man."

the thirty-one men he watched die . . . No women have been asphyxiated in Mississippi. Three women were gassed in San Quentin, including Barbara Graham, heroine of the book and movie, *I Want to Live,* who did in fact enter the chamber uttering the words later immortalized by Susan Hayward: "Good people are always so sure they're right." They were, however, not her final words: on being told by her death watch jailer Joe Ferretti, "Take a deep breath and it'll go easier," she snapped back, "How in the hell would you know?"

"fooling with thangs" . . . If any generality can be said to apply to executioners, it would be a strong work ethic, generally born of hard times. Bruce was fortunate in that his work as an executioner did not cost him other jobs. Henry "the Hangman" Meyer, an elderly carpenter who became the Louisiana hangman in his mid-fifties, was unhirable after he took the hangman's job. The above-mentioned Canadian hangman Arthur Ellis lost a prized waiter's job at the Toronto Yacht Club when a Mountie (who had either served at or attended one of Ellis's hangings) told the maître d' he would prefer not to be served his meals by a "common hangman." The irony was bitter for Ellis, who was often called upon in provinces when the local Mounties (normally entrusted with executions) refused to do the job. Ellis's career on the gallows ended in 1935 when he decapitated Thomasino Sarao, who prison officials had assured Ellis weighed 145 pounds. Sarao in fact weighed 187 pounds and the extra few feet of drop led to her losing her head. Ellis went to his death in 1938 still a staunch supporter of the penalty, though he supported the electric chair as being "faster and less dangerous" to "those who performed the execution."

gala affairs . . . One of the last American public hangings, in Owensboro, Kentucky, in 1936, drew a crowd estimated at fifteen to twenty thousand. A reporter covering the event for a Tennessee newspaper was struck by the festiveness, and reminded his readers that the word *gala* is derived from "gallows." If that etymology is valid (the Websters don't agree), Kentucky and nearby Tennessee would be the place to verify it: in a double hanging of John Hall and Burrell Smithy in Murfreesboro, 15,000 onlookers were treated to a barbecue, and whiskey, beer, and sarsaparilla were served at a makeshift amphitheater. In the ensuing drunkenness, a grandstand erected for viewing the event collapsed, causing a fatal stampede. Mississippi and Louisiana hanged semi-publicly until 1938 and 1940, respectively. Louisiana's last hanging, in a Monroe prison, of four murderers who'd gone on a rampage after escaping from an Arkansas prison, was broadcast by the local radio station.

Putting all executions behind Parchman's walls . . . Other than Louisiana, which had its own provincial issues, Mississippi was the last state to conduct executions in a central prison, a trend begun in the 1830s by Pennsylvania and New York legislators embarrassed by the scenes of mass drunkenness, rioting, and looting that invariably accompanied public hangings.

no shame in handing a corpse like this back to the family . . . The asphyxiation may have had an unintended effect on the Gallego family. His son, Gerald Jr., nine years old when his father died, was sent to a youth correctional facility four years later for committing a "lewd and lascivious sexual act on a six-year-old girl." He went on to become a "notorious sex-slave killer" (at least as dubbed by the *Las Vegas Sun,* 2/24/98) who torture-murdered couples kidnapped from malls. He achieved a rare distinction when he became the first man post-*Gregg* to be sentenced to the gas chambers of both California and Nevada. After repeated stays, including one caused by a district attorney's office's failure to mail a court document in on time (the secretary said she wanted to "save the office a few bucks and chose regular mail instead of Overnight"), Gallego managed to outlast the gas both in Nevada and San Quentin, where lethal injections are now used exclusively. He is currently awaiting sentence in both states. "What a Way to Go," a fact-filled article by Irwin Moskowitz in the November 1962 issue of *Inside Detective,* makes the excellent point, while reviewing the forms adopted, that alternative execution methods arose partly because the United States returns the corpses of the condemned to next of kin, a practice that may be unique worldwide.

sold vegetables at a roadside stand . . . Historically, executioners have often doubled as butchers, chefs, or keepers of the larder. Medieval executioners' responsibilities often included disposal of animal carcasses. In Paris, the axman was given license to slaughter pigs, a prized position, though he was also charged with cleaning the local sewer. From a typical early "protocol," this for the

1541 amputation of a hand and consequent execution of Edmund Knevet, who had struck a servant of the Earl of Surrey on the tennis court of the King's House, we see that seven of the ten domestics employed in the killing came from the king's kitchen: "3. The master cooke of the king with the knife. 4. The sergeant of the larder to set the knife right on the joint. [The hand that had struck the Earl of Surrey's servant was amputated before Knevet was decapitated.] 5. The sergeant farrier [responsible not only for shoeing horses but for slaughtering them for table], with his searing-irons to sear the veins. 6. The sergeant of the poultry, with a cocke, which cocke should have his head smitten off on the same blocke [as Knevet's head]. . . . 8. The yeoman of the scullery with a pan to heate the yrons; a chafer of water to cool the ends of the yrons and fourmes for all the officers to set their stuffe on. 9. The sergeant of the cellar with wine, ale, and beere. 10. The yeoman of the eury . . . with bason, eure, and towels." Manuel Noriega's chef, on at least one occasion, served as torturer-executioner. As reported in the *Village Voice* in 1988, he gutted a "subversive journalist" with a ten-inch Hoffritz chef's knife after a week of torture in a basement cell, then decapitated him with the same knife. The executioner was sometimes paid partially in food. A Highlands colloquialism, "mercks and perks," meaning wealth that comes in dribs and drabs, derives from the axman's traditional pay: a few favors—such as a bed for the night previous to and after the beheading, the cleaning of his clothes, and a few handfuls (or mercks) of grain. The sixteenth-century executioner at Kaufbeuren, Germany, received four florins for decapitation, burning, or hanging; only two florins for crushing with the wheel; but fifteen florins "for the customary meal after such an execution."

the occasional piece of anonymous hate mail . . . The electrocutioner Robert Elliot was threatened in hate mail prior to most executions. He also received many requests for autographs and get-well cards whenever the newspapers carried a report of his illness.

working for "blood money" . . . It can be argued that the executioner's pay constitutes not so much "blood money" as a sublimation of the restitution that victims' families historically received, sometimes in lieu of the execution itself: "Even today in Khartoum, Sudan, the current interpretation of the 1350-year-old Islamic *sharia*, an ethical, criminal, and civil code, allows murderers to escape the death penalty by paying to the victim's family what is, called *diyyah* or blood money in the amount of one hundred camels or a negotiated equivalent which has recently been legally standardized at close to $17,000" (Anne Butler, *Dying to Tell,* Lafayette: University of Lousiana, 1992, p. 88). Western sums that might qualify as "blood money" were the tips that nobility gave the swordsman to take the head off cleanly. The sword, generally a large saber, was used only for nobility and tradesmen. Commoners were dispatched with an ax, which is apparently a poor killing tool. The job was by tradition hereditary, and the practice was for the executioner's son to complete any bungled job. . . . *or thought him notorious* . . . The dishonor of the profession, where it exists, seems as old as the separation between judgment—an exercise of wisdom—and execution of sentence, reviled by its association with retribution and violence. Aristotle, *Politics*, Book Six, Part One, is an early authority: "The difficulty of this office arises out of the odium that is attached to it; no one will undertake it unless great profits are to be made, and any one who does is loath to execute the law." Medieval executioners, who were often dismemberers and torturers, were regarded and consulted as despised members of the healing profession, as their knowledge of the body was superior.

got particularly quiet when he drank . . . The nineteenth-century Canadian hangman John Robert Radclive, who hanged anywhere from 132 to over 200 men, drank heavily. He adopted the often faulty method (used in Utah, Pennsylvania, Montana, Illinois, etc.) known variously as a jerk'em, jerk gallows, or jerk-'em-up gallows, in which the condemned was yanked up suddenly by hand or by a pulley mechanism, a method thought more apt to sever the spinal column than a simple drop. The jerk-'em-up failed for Radclive in the execution of a Reginald Bichall, who strangled for almost twenty minutes. Radclive's drinking became heavier after hanging Bichall, and he became a hopeless drunk after a particularly harrowing execution in St. Scholastique, Quebec, when the condemned suffered heart failure and fell dead in his arms on the scaffold. The local sheriff ordered the man hanged regardless, and Radclive had no choice but to comply.

"I used to say to condemned persons as I beckoned with my hand, 'Come with me,'" Radclive wrote at the end of his life. "Now at night, when I lie down, I start up with a roar as victim after victim comes up before me. I can see them on the trap, waiting a second before they face their Maker. They haunt me and taunt me until I am nearly crazy with an unearthly fear. I am two hundred times a murderer, but I won't kill another man." Radclive drank himself to death in 1912.

CHAPTER FOUR

carried out by accident despite a last-minute stay . . . The legend about Justice Warren's failed stay does not appear to be valid, though it was printed in several Mississippi and Tennessee newspapers before hitting the national press, or at least *The Nation.*

a strange-looking procedure . . . In California, the dust-down team never touched the corpse inside the chamber. Instead, they went in with garden sprayers filled with concentrated ammonia.

"Fucking guy ain't dead!" . . . The corpse of John Brown looked lifelike hours after he was cut down from the Charlestown gallows, though he took a full thirty-five minutes to die. Doctors at first wouldn't sign the death certificate, for fear a "galvanic battery" would bring him back to life. This appears to have been a common misconception of the era: men hanged at Tyburn and Lancaster in the mid-nineteenth century were normally left on the gallows for a half hour to an hour to ensure no chance of revivification. The above-cited nineteenth-

century Tennessee condemned men John Hall and Burrell Smith convinced not only Murfreesboro doctors but attending lawmen to attempt their resurrection with a galvanic battery, but the experiment failed.

Fears of revivification were, naturally, exacerbated with the move to the electric chair in New York. At a July 15, 1889, hearing on the new method, a Mr. Charles Tupper, restaurant owner, described how "his dog, Dash, a Scotch Collie and St. Bernard mix, was shocked by a fallen telegraph wire that had fallen over a power line. Dash appeared dead and . . . was placed in a hole 'to draw electricity from him' at 4 pm. By 10 pm, he started moving and by 1 to 2 am he was up and about. Dash was a star exhibit at the hearings when he was brought in. Another witness . . . Alfred West . . . was struck by lightning in 1880 or 1881. He was revived . . . when his feet were placed in warm water and his rescuer pulled on Mr. West's toes with one hand and milked a cow with the other." ("Theories of the Causes of Death from Electricity in the Late Nineteenth Century," *Medical Instrumentation* 9, 6 [Nov.–Dec.], 1975.)

The August 1945 issue of *Front Page Detective* carries an account of "two Russian scientists," Dr. Vladimir Negovsky and his student, Arkady Makarychev, who "restored to life 50 Soviet soldiers found on the battlefield in a state of clinical death." The method reported was an injection of "blood enriched with glucose and an adrenaline content," and "artificial respiration applied through a tube directly into the windpipe. The dead men came to life!"

Margie Velma Barfield, the first woman executed post-*Gregg*, unwittingly rekindled the electrical-revivification controversy with her last-minute efforts to atone for her crimes—she'd poisoned her mother, husband, and three others with roach powder. Barfield, after long ministering with the Reverend Billy Graham (who brought his daughter to death row to meet her), willed her kidneys to the North Carolina Baptist Hospital in Winston-Salem prior to her 1984 execution by lethal injection in Raleigh's Central Prison. The kidneys would be harvestable, however, only if blood was flowing in her body, and that meant her heart would have to be restarted. The few remaining members of her family, not surprisingly, had no wish to see her revivified, and refused to sign a form allowing the use of heart stimulants. Central Prison authorities, fearing "a scene in which Barfield would sit up in the chamber," also refused to release the body until the heart had stopped for five minutes. Shortly after her execution, a local hospital, fearing publicity, reneged on its promise to provide an operating room for the harvest, and the paramedics and doctor engaged to take the corpse to the harvest decided on a Winston-Salem hospital. In the back of an ambulance, they pumped oxygen into her lungs throughout the 100-mile drive to the hospital. Her cheeks showed occasional pinkness, but her heart never beat again, and doctors gave up on the kidney harvest well before they arrived. In Winston-Salem, however, they were able to take her eyes, as well as some bone and skin.

CHAPTER FIVE

the narrowest kennel runs . . . H-O-R-S-E, the shooting-challenge game featured in mid-1990s McDonald's ads, began in the death row yards of the Nevada State Prison in Carson City. The inmates, not allowed basketballs, challenged each other across the razor-wire fence with trick shots of rolled-up socks and underwear. It being Nevada, the game became a form of gambling. Typical stakes were a hundred push-ups for the loser. Nevada's death row prisoners now have balls and hoops, but the wire between kennel runs is still covered with underwear that once cost someone a letter.

A man called Top Cat . . . The names or nicknames of several of the noncondemned MSU prisoners have been altered.

he had been too drunk once to carry out an execution . . . The rumor of the executioner's drunken inability to run the chamber (Bruce was not mentioned by name) was represented as fact in one of Jimmy Lee Gray's motions before the Fifth Circuit. No substantiating evidence was provided.

you drew lines in the dirt and shot anyone who crossed them . . .
The original "deadline" for newspaper reporters was established by
the warden of Sing Sing on July 7, 1891. The prison was holding its
first electrocutions—a four-man execution—and the New York
City reporters from down the Hudson came up by the dozens, all
demanding to be let in. The warden drew a line outside the wall and
gave armed guards orders to shoot any reporter who crossed.

The "deadline" set two other precedents. By way of reporting
the deaths of the four men inside to the assembled press, flags of
different colors were waved from the death house door as each man
met his end. A black flag was waved for the one black man. Until
Furman, the waving of a black flag from the door of the death
house was the signal in a number of states that sentence had been
carried out. The second, more crucial, precedent was the lengths wardens would travel to ensure secrecy: while there had been twenty-five
witnesses to Kemmler's electrocution, the number was limited at Sing
Sing; vows of secrecy were sworn, and reporters were excluded. The
only detail that seems to have emerged from this historically important quadruple execution was that the black prisoner had received
an extended jolt of electricity—because his skull was thought to be
"thicker" than the others'. (Amateur, ad hominem determinations of
proper current were common: Jimmy Thompson, Mississippi's first
electrocutioner, gave rapists "extra juice," as he deemed them "a
special kind"; women, conversely, were given lighter shocks.) By the
next Sing Sing electrocution, however, the newspaper accounts were
as graphic as they were numerous: how the condemned had to be
carried in a state of near unconsciousness to the chair, where he was
either mistakenly or intentionally given four shocks instead of
three, how his skin had roasted, his left eyeball had popped out and
the "aqueous humor" had run down his cheek, etc. Such reporting
became a commonplace of early electrocutions.

CHAPTER SIX

drench his shirt . . . Cabana remembers Vietnam vets at Parchman
who "seemed like classic studies in Post-Traumatic-Stress Disorder,

and probably didn't belong at the end of a state-issued gun." One K-Nine unit member fell into an eerie mantra whenever they caught the trail of an escapee: "Let's get some ears, get some ears . . ."

this sickly magnetism . . . The extreme sexual allure of serial sex killers like the Boston Strangler, Ted Bundy, or Richard Ramirez is the hot-button gossip of wardens at American Correctional Association conventions.

Missouri's chamber leaked badly . . . Missouri switched to lethal injection in 1988 but failed to get a death chamber installed before their first execution, which was of Tiny Mercer, on January 6, 1989. He was killed by lethal injection on a gurney put inside the gas chamber, its door and two chairs removed. Tiny, who had lost a lot of weight and was a former drug addict, had to be excavated, or "cut down," in the groin area to find a usable vein. He was the first man executed by the Leuchter Machine, a device made by the execution technician Fred Leuchter that automated lethal injections and was activated by one of three keys turned in a metal box. The machine resulted in several botched executions, including that of John Wayne Gacy in Illinois, but Leuchter remained the nation's major execution technician until his collaboration with neo-Nazi groups (trying to disprove the "myth of the Holocaust") was made known to corrections officials nationwide by an Alabama bureaucrat who had become antagonized by Leuchter's relentless profiteering and critiques of any methods other than his own. When Maryland canceled its order for one of his lethal-injection machines, he tried to sell it in a local classified magazine. "FOR SALE," read the copy, sandwiched between ads for Beatles cards and scuba gear: "EXECUTION DEVICE. $10,000."

"moved—as if by some unseen force" . . . from *Death at Midnight: The Confession of an Executioner*, by Donald A. Cabana (Boston: Northeastern University Press, 1996).

a bit of English . . . The figure, which in fact seems extremely high, doubtless included a ballpark guess at parts and labor. Hocutt and van Landingham, who did the repainting, reswabbing, testing, etc., were salaried officers, however; the rare replacement of meaningful cost was the door's large oval gasket, which had to be handmade. Eaton Industries had stopped making parts during the moratorium, and all remaining inventory (as well as plans and blueprints) had been lost to a factory flood in the early 1970s. Gas chambers, nonetheless, have the priciest start-up costs of any form of execution. Fred Leuchter's chamber, for example, went for $200,000, while his electric chair sold for only $35,000.

put a man to sleep or . . . *kill him* . . . "The Big Sleep" was a phrase inadvertently coined in the 1930s by San Quentin's Warden Duffy, hoping to assure the first man he asphyxiated that it was "like going to sleep." The slang evolved as Duffy's words passed down death row. Prison records of subsequent asphyxiations indicated this was not the case. Nonetheless, in 1992, San Quentin's warden Dan Vasquez advanced the "big sleep" theory when asked about the difference between the gas chamber and lethal injection by David Mason, a man he subsequently asphyxiated. "Both methods," he said, "are equally humane," adding that the gas would make Mason unconscious "in two to four seconds."

would not be the man he first executed . . . Pruett was extradited to Arkansas, where he remains, as of this writing, on death row.

maybe taking on a younger man . . . The oldest executioner known is a headsman named Reichard, who at the age of eighty beheaded a man in Regensburg, Germany, in 1931.

his crime and particulars bore an unfortunate resemblance . . . Though the Court's decision led to a huge spate of 1987 executions, Warren McCleskey himself escaped execution for another four years.

framed and forced to confess by the police . . . A condemned man in San Quentin, who contributed a dozen entries to the Web site "Au Revoir Les Inmates," passed on death row's verdict of O. J. Simpson halfway through his trial: "a guilty man, framed by the LAPD." In another entry, he explained that the row's least favorite movie was *Groundhog Day,* because it reminded them too much of their own experience of waking up every morning to the exact same routine.

a decent game of chess . . . Average chess strength among U.S. inmates is high. The caliber of condemned players is impossible to determine, however, as they are not allowed to compete—even in postal chess. Charles Manson once held a United States Chess Federation rating of 2,150—only fifty points, or a half dozen or so significant victories, short of the master rating.

CHAPTER SEVEN

had been right, too, wanting to cut this visit short . . . The point was driven home at a 1989 Georgia execution. A woman who had married the condemned man a week prior to the execution collapsed at his feet and wouldn't let go of his leg when he was summoned for his last mile. The death watch guard, a woman who was later severely reprimanded by the Georgia DOC, was unwilling to intercede. Two additional guards were called in to pry the condemned free from his wife. In San Quentin, which set the standard for gas chamber protocols in the 1930s, visitation ends twenty-four to forty-eight hours before the execution date.

absolve the warden . . . Louis "Big Red" Nelson, warden of San Quentin in the mid-1950s, remembered a poignant example of this absolution. Minutes before a double execution on April 6, 1956, one of the men, Robert Pierce, managed to slit his throat with a sliver of mirror he'd secreted into the chamber in his mouth. He missed the carotid artery but bled copiously as the death watch jailers pulled him from the chamber. Wrapping his blue workshirt around his neck, they dressed him in a fresh white shirt, then dragged him back. While he sat, bleeding to death, "[h]e called [the witnesses] every filthy name in the book," said Nelson. But he took time to nod toward the prison officials in his view to let them know he didn't hold them responsible.

CHAPTER EIGHT

What if they were allergic to hydrogen-cyanide? . . . Luis José Monge, the last man asphyxiated pre-*Furman,* asked the physician attending his 1967 Colorado execution if the gas would bother his asthma. "Not for long," he was told.

it should be as public as possible . . . Executions in Guatemala have been broadcast on local TV since 1996. Two U.S. Spanish-language networks, Univision and Telemundo, ran footage of a June 29, 2000, execution on U.S. channels.

EPILOGUE

where they think *they're going* . . . Leong Fook, hanged in San Quentin in 1929, was allowed to wear neither his eyeglasses nor his false teeth to the gallows, though he insisted he would need them where he was going. It was finally agreed they would be placed with him in his coffin. Boogie Woogie La Fontaine, a twenty-year-old Mississippian executed by Bruce, had a pack of Camels in his rolled-up T-shirt sleeve as he entered the chamber. "Where I'm going," he explained, "there may not be any cigarettes." Jesse Bishop, asphyxiated in Nevada, October 22, 1979, put thirty-five cents in his pocket, and told the warden to tell his family he'd call them when he got to wherever he was going. Several mentally retarded or brain-damaged inmates have left the dessert of last meals in their cells, explaining they would come back to eat them later.

it puts all heaven in an uproar . . . The 2000 Harris Poll on the penalty showed that, on average, Americans both for and against the penalty believe 10 percent of the condemned are innocent.